A.I. Meets AEC:

How to Harness Artificial Intelligence to Supercharge your Firm

PSMJ | Resources, Inc.®

First published in 2024 by:
PSMJ Resources, Inc.
10 Midland Avenue
Newton, MA 02458
Phone: 617-965-0055
Email: customerservice@psmj.com
www.psmj.com

ISBN: 978-1-55538-005-2

Table of CONTENTS

Table of CONTENTS

Introduction
The A.I. Revolution

The promise of artificial intelligence (A.I.) has hovered over humanity for decades, both exciting and frightening us. It has been the subject of sci-fi movies and dense technical research papers for as long as most of us have been alive. And in 2023, it burst through the front door and planted itself on our work stations, laptops, phones and personal devices.

The year 2023 will likely be remembered as the birth of mainstream A.I., not just in the architecture, engineering and construction (AEC) industry, but in all areas of business and life. That is, unless something more dramatic happens with A.I. in future years that will make everyone forget about what happened at the beginning. This is not unlikely.

The purpose of this book is to provide non-technical leaders and other professionals in the AEC field with a resource to better understand and use A.I. It traces the evolution of this monumental advancement in how we use technology – and how technology serves us. It shows how A.I. is being used in AEC firms and everyday life in 2024, and how its use is likely to grow and change in the future. Most importantly, it illustrates how AEC firms can capitalize on this earthshaking revolution in the ways that we run our business, provide our services and deliver our projects.

The book seeks to answer these primary questions:

- What is A.I., and what isn't A.I.?

- What are the primary components of A.I. and how do they interrelate?

- How is A.I. being used in AEC firms (and in everyday life) today?
- How can firm leaders manage the change that A.I. has already brought to the industry and better position their organization for the lightning-fast pace of change?
- What best practices can firms use to integrate and optimize A.I. and other related and unrelated evolving technologies?
- What's coming next?

The book is the culmination of a vast amount of research on the topic, both general and industry-specific, along with the contributions of several technology experts who are quoted throughout. It also relied on some of the A.I. tools discussed in the book, including ChatGPT, DALL-E and Copilot.

PSMJ hopes you find this book informative, entertaining and ultimately useful in your understanding and use of A.I. and technology in general. Thank you for reading.

Jerry Guerra

March 2024

ACKNOWLEDGEMENTS

The author thanks the following subject matter experts and secondary research sources for their contributions to this book.

Brian Burnett, PSMJ

Bruce Bartolf, Forestay Advisors

Dave Burstein, PSMJ

David Mulholland, VHB

Frank Lazaro, AE Marketer

Frank Stasiowski, PSMJ

Gregory Hart, PSMJ

KP Reddy, Shadow Ventures

Justin Hoff, Martin Gardner Architecture

Lucas Hayden, Unanet

Lud Eng, Stratus

Mike Boisvert, Verdantas

Scott Steiding, Productive Pursuits

Steve Bennett, Taylor Design

Chapter

01

WHAT IS (AND ISN'T) ARTIFICIAL INTELLIGENCE

Of the many mysteries surrounding the recent eruption of artificial intelligence (A.I.) as an everyday part of human society, the most mystifying may be this – almost no one can say exactly what A.I. is. At least not in a way that is easily comprehensible to the average non-technical person.

Is it a technology? Is it a process? Is it a strategy? A product? A theory?

This is a problem. When you hear something identified as artificial intelligence (A.I.), it often isn't. In many cases, A.I. has become a catchall phrase that fails to distinguish it from related and unrelated technology tools and systems. It is often confused with or portrayed as synonymous with emerging technologies such as machine learning, natural language processing (NLP), alternate and virtual reality (AR/VR), 3D modeling and so on. Some of these technologies overlap with A.I., and some aren't A.I. at all.

So what is A.I. really? And what isn't A.I.?

This chapter defines the elements of A.I. and describes how they interconnect. It also includes a section about technologies and activities that are often confused with A.I., but do not necessarily meet the standards of A.I.

DEFINITIONS OF A.I.

Discussions about "A.I." – and it seems that most conversations and content produced today are about A.I. – often overlook the nuances of the technologies, systems and management practices that co-exist with or fall under the A.I. umbrella. This disconnect detracts from our understanding of how A.I. can best suit our lives and advance our professional interests, especially for those of us who are not professional technology experts.

"The linguistics around A.I. are very obtuse, to the point that the term 'A.I.' doesn't mean much," says Bruce Bartolf, Senior Advisor for Forestay Advisors, LLC, and the former Chief Technology Officer (CTO) for Gensler. "People wrap machine learning into A.I. Yes, it's a branch of A.I., but it's not A.I. Algorithms, machine learning, and neural networks are all types of A.I. How do you communicate these subtle, but important differences? Marketing people throw the term A.I. around because it's fashionable, but that starts to erase the meaning. We need to come up with better terminology so people can understand what A.I. really is."

So what is it? One of the simplest and clearest definitions of A.I. comes from Encyclopedia Britannica: "Artificial intelligence (A.I.) [is] the ability of a digital computer or computer-controlled robot to perform tasks commonly associated with intelligent beings."

Great! It is computers doing things that humans can do. But how?

Diving more deeply into the subject, the generative A.I. program ChatGPT offered this when asked what A.I. is: "A.I. refers to the simulation of human intelligence in machines that are programmed to think and mimic human-like cognitive functions. A.I. systems are designed to perform tasks that typically require human intelligence, such as visual perception, speech recognition, decision-making, and language translation."

Kind of a long way of saying the same thing, but with a bit more detail.

Online training course provider Simplilearn offers a more complete definition that includes a key word omitted by the others – "data." Simplilearn says, "Artificial Intelligence is the process of building intelligent machines from vast volumes of data. Systems learn from past learning and experiences and perform human-like tasks. It enhances the speed, precision, and effectiveness of human efforts. A.I. uses complex algorithms and methods to build machines that can make decisions on their own. Machine Learning and Deep Learning form the core of Artificial Intelligence."

Note that this definition calls A.I. a process. ChatGPT called it a system. Encyclopedia Brittanica, an ability. So which is it?

The answer is, A.I. represents all three. And more. But what A.I. is not is a technology.

Here is another way to say it. **A.I. is the result of machines performing tasks that humans do, albeit more quickly, accurately and efficiently.**

" A.I. within the AEC industry is coming," says David Mulholland, Chief Technology Officer (CTO) at VHB, a 2,000-person civil engineering, consulting, and design firm. "Today, A.I. is often a buzz-term for full automation with limited human interaction. The machine learning aspect seems to generally be the iterative part that everyone is referring to as A.I. While A.I. is coming rapidly, there appears to be three stages. First is the base framework for the models. This includes the stagnant and dynamic data that can be fed into a learning model. The second stage is the machine learning aspect. Use machine learning to help make informed decisions or analysis. Generally, the third stage is the automation of A.I. Leveraging A.I. tools to automate the analysis process and provide automated and sometimes predictive analytics."

So to understand what A.I. is, we need to become more familiar with the components that allow us to achieve A.I. This includes machine learning and deep learning.

MACHINE LEARNING VS. DEEP LEARNING

Machine Learning is exactly what it sounds like. It is when machines exposed to certain types of data can "learn" from it to perform functions better, faster and more accurately. IBM defines machine learning as "a branch of artificial intelligence (A.I.) and computer science [that] focuses on the use of data and algorithms to imitate the way that humans learn, gradually improving its accuracy."

In many cases, when people use the term " A.I.," they're actually talking about machine learning.

"Machine learning is an algorithm that analyzes data and learns what's been done," says Forestay's Bartolf. "A process is considered A.I. when it uses the knowledge that machine learning has gained and extrapolates it to do new things."

Deep Learning is a subset of – or perhaps more accurately, an evolution of – machine learning. It "is essentially a neural network with three or more layers," says IBM in its article "What is Deep Learning?" on ibm.com. "These neural networks attempt to simulate the behavior of the human brain – albeit far from matching its ability – allowing it to 'learn' from large amounts of data. While a neural network with a single layer can still make approximate predictions, additional hidden layers can help to optimize and refine for accuracy. Deep learning algorithms are most strongly associated with human-level A.I."

This explanation of deep learning begs the question:
What is a neural network?

Neural Networks use numbers and mathematical information for data processing. With data nodes that resemble neurons and synapses, they emulate the functioning of the human brain.

Taking this comparison further, deep learning teaches computers to process data in a way that is inspired by the human brain. It can recognize complex patterns in not only text, as Chat GPT does, but also with images, sounds and other kinds of data. It can then quickly and accurately produce relevant content, insight and inferences.

Though machine learning and deep learning are both based in algorithms that use data to learn, the key difference is how they process and learn from it. In a November 2023 article, "Deep Learning vs. Machine Learning," Software company Zendesk explains, "While basic machine learning models do become progressively better at performing their specific functions as they take in new data, they still need some human intervention. If an A.I. algorithm returns an inaccurate prediction, an engineer has to step in and make adjustments."

Zendesk continues, "With a deep learning model, an algorithm can determine whether or not a prediction is accurate through its own neural network – minimal to no human help is required. A deep learning model is able to learn through its own method of computing – a technique that makes it seem like it has its own brain." So a major difference is in how much humans need to get involved for the algorithms to do their job.

Simply put, the most basic machine learning digests data, then performs the function it was designed to perform. Deep learning digests data, but can take the "knowledge" it gains to generate its own data, and then enhance that data as it continues to "learn."

"Machine Learning is best for well-defined tasks with structured and labeled data," says an article on the Amazon Web Services (AWS) website. "Deep learning is best for complex tasks that require machines to make sense of unstructured data. Machine learning solves problems through statistics and mathematics. Deep learning combines statistics and mathematics with neural network architecture."

Another major difference is that machine learning consists of thousands of data points, while deep learning uses millions. Machine learning functions perfectly well using small datasets, but deep learning requires much larger amounts of data to understand and perform in ways that exceed basic machine learning's abilities.

Programming for machine learning is focused and explicit, while deep learning requires layers of neural networks. And deep learning algorithms require far more time to train than machine learning – in some cases hours instead of days. Finally, because they're more complex and require larger datasets, deep learning models demand more storage and computation power than machine learning models.

NATURAL LANGUAGE PROCESSING

A.I.'s recent expansion into previously untapped areas owes a debt to the multiple existing technologies whose continuing evolution allowed it to break through barriers and go beyond prior limitations. **Natural Language Processing (NLP)** is one of these.

NLP is a branch of artificial intelligence (A.I.) that enables computers to comprehend, generate, and manipulate human language. It combines machine learning with computational linguistics, statistics, and deep learning models to process human language from voice or text data. Its superpower is the ability to analyze data with natural language text or voice.

NLP drives many of the everyday technology tools that we've come to rely on in the past decade or so. Inspired by an article on tableau.com ("Eight Natural Language Processing Examples"), here are some programs and functions that rely on NLP:

Email filters are an early and simple example of NLP. Spam filters that recognize suspicious words and patterns have been at work for years, but the tools are becoming even more sophisticated.

Smart assistants recognize speech patterns using voice recognition, then infer meaning and provide a useful response.

Search results. Search engines use NLP to interpret the intent of the search query, rather than relying exclusively on the explicit data it receives.

Predictive text. Autocorrect, autocomplete and predictive texts learn our habits to customize the convenience they provide (or, too often, try and fail to provide).

Language translators. NLP allows language translators to deliver more accurate results because it can "learn" important factors such as word organization and context.

Automated phone systems. NLP is largely responsible for the automated phone systems that attempt to mimic human attendants (but so often frustrate us). These digital systems "learn" from recording phone calls and use their newfound knowledge to attempt to improve. (Editorial note: As amazing as this technology is, I still find myself angrily yelling "representative" at these systems far too often.)

Data Analysis and Text Analytics. In both cases, NLP's ability to understand and interpret human language allows faster and more robust output. For example, with NLP's help, A.I. bots can generally interpret text, content, and images to allow for a pdf file "chat." You can ask one of the many "chat with pdf" tools available to examine a pdf document and tell you what's in it (within the confines of your information request) without having to search for specific words or phrases.

In text analytics, NLP allows an A.I. program to interact with unstructured data in a conversational way, and it can also help manage large volumes of survey results faster and more accurately than humans. For example, a tool using NLP can scour customer interactions, such as social media comments or brand name mentions, and analyze these interactions to help determine how a marketing campaign is performing.

OTHER A.I.-RELATED DEFINITIONS

Certain words and phrases continually arise in discussions of A.I. Here is a brief glossary of some of these important words and phrases.

Generative A.I. Introduced in the 1960s, generative A.I. (Gen A.I.) produces various types of content, including text, imagery, audio and synthetic data. It advanced beyond simple chatbots with the introduction of generative adversarial networks (GANs) in 2014, and grew more mainstream with the release of more user-friendly tools such as ChatGPT in 2022.

Generative adversarial networks (GANs). An example of deep learning architecture, GANs train on a given dataset, pitting two neural networks against each other to achieve the most authentic output. GANs consist of a generator neural network that creates new data and a discriminator neural network that judges the accuracy of the generator's output. As this dynamic progresses, both networks "learn" how to perform their assigned function better until they achieve equilibrium – where the discriminator can only identify false data half the time. The goal of this virtual war of wills is to create artificial data that could be easily mistaken as real.

Generative Pre-Trained Transformer Models (GPTs). Another type of generative A.I. that can be used to generate text, translate languages, and write different kinds of creative content. It's the "GPT" in ChatGPT.

Large Language Model (LLM). Technology company Nvidia describes LLMs as "deep learning algorithms that can recognize, summarize, translate, predict, and generate content using very large datasets." The evolution of LLMs was a crucial step in the progression of A.I. from background character actor to frontline feature player.

Optical Character Recognition (OCR) is the technology that recognizes text in documents and images.

Machine Vision. This is when a computer can "see." Tech company Intel says, "Machine vision combines smart cameras, industrial PCs, and deep learning models to create intelligent systems for advanced industrial automation."

Computer vision uses pattern recognition and deep learning to identify content in images and videos. It processes, analyzes and obtains knowledge about images and videos, allowing A.I. tools to interpret surroundings in real-time. Machine vision is a subset of computer vision. Zebra Networks says, "Computer vision is a broader term involving the automation of image analysis without relying on human eyes," while "machine vision uses computer vision in industrial and practical applications."

The full breadth of A.I. involves many other technologies, systems, tools and processes, each one of which can take you down a new rabbit hole. We've included most of the basic foundational elements relevant to A.I. in the AEC industry, but this is in no way a full representation of all that exists in this fast-evolving field.

TWO TYPES OF A.I.: “WEAK” VS. “STRONG”

A.I. comes in two major categories, narrow or weak A.I. and general or strong A.I. The latter currently exists largely in theory.

Narrow or Weak A.I. is designed and trained for a specific task or set of tasks. It operates within a limited context and cannot perform tasks beyond its predefined capabilities. Virtual assistant programs like the iPhone’s Siri are examples of Weak A.I. Other models include image recognition software, and recommendation algorithms for social media and streaming networks.

General or Strong A.I. refers to machines with the ability to apply intelligence to any problem, rather than being limited to a specific task. It is a more advanced form of A.I. that can understand, learn, and apply knowledge across diverse domains, similar to human intelligence. Most commonly called “artificial general intelligence” or AGI, it exists as more of a theoretical concept and has not been fully realized yet.

Shane Legg, Google DeepMind’s co-founder and chief AGI scientist, told *Time* that he expects AGI to be developed by 2028. Other A.I. experts predict that AGI will become a reality within 5 to 20 years.

Perhaps as a way to temper expectations, *Time* also noted that Herbert Simon, who won a Nobel Prize for economics and the Turing Award for computer science, predicted that “machines will be capable, within 20 years, of doing any work that a man can do.” That was in 1960.

Clearly, nothing is guaranteed. As advanced as the recent (and continuing) chatbot and other machine-learning-based technologies have become, the consensus among experts is that they still fall short of AGI.

In 2024, AGI exists only in works of fiction. Arnold Schwarzenegger's character in The Terminator is an example of AGI. Though it had a single focus – kill Sarah Connors – it was able to "understand, learn and apply knowledge across diverse domains." The Terminator is a character of fiction – for now. As A.I.-contributing technologies advance, we grow closer than ever to true AGI.

Below is a graphic from technology information and advice website Lifewire (lifewire.com) that explains the differences between the weak/narrow and strong/general A.I.

Weak or Narrow A.I.	Strong or General A.I.
Performs one specific task	Can perform any ask a human can
Programmed for a certain purpose	Learns how to perform new skills
Learns how to perform tasks faster	Uses creativity to solve problems
No self-awareness	Potentially sentient

A third category is sometimes included in this group – Artificial Superintelligence (ASI). This refers to a hypothetical form of A.I. that surpasses human intelligence in every aspect (presumably represented by Arnold Schwarzenegger's "emotional" character in Terminator 2). Discussions of ASI revolve around the obvious ethical, safety and existential questions of building such a machine. And for the purposes of this book, that's enough about that.

SO WHAT IS & ISN'T A.I.?

In a 2020 article, ancient by technology standards but still pertinent, Mike Kaput of the Marketing Artificial Intelligence Institute wrote, "Artificial intelligence is a broad term. It encompasses a lot of different technologies. Some of these technologies are called machine learning, natural language processing (NLP), natural language generation (NLG), deep learning, and neural networks."

He continues, "Some commentators treat these concepts as interchangeable. They're not. They refer to specific, but related, technologies. The term 'artificial intelligence' or 'A.I.' is an umbrella term to describe this suite of tools. Kind of like how the term 'analytics' or 'business intelligence' encompasses many technologies. Understanding this helps a lot of people start to wrap their heads around A.I. You pull back the curtain and see individual tools with real capabilities, not some singular magic technology that can do everything."

And there it is. A.I. as it exists today encompasses a range of technologies that simulate or replicate human intelligence to perform specific tasks. While it has made substantial strides in recent years, with 2023 being the pinnacle thus far, it is critical to understand A.I.'s limitations and to differentiate between the narrow degree of capabilities it currently offers and the theoretical concepts of general or "strong" A.I.

SIDE NOTES: " A.I. IS NOT A TECHNOLOGY"

A 2018 article in *Forbes* argues that, "Artificial intelligence is not a technology." Author Kathleen Walch writes, "Asking the question whether or not some particular technology is or isn't A.I. is missing the point. Artificial intelligence is the journey. It's the quest for the intelligent machine. All the technologies we've developed on the route to that quest are things that are individually useful, but all together, have not yet gotten us to the goal. This is why it's important to understand that artificial intelligence is not a technology, in much the same way that the Space Race is not a technology."

Walch adds, "The term artificial intelligence was coined in 1956 at a Dartmouth University summer conference. John McCarthy, celebrated A.I. researcher who pulled together the conference himself, said that A.I. doesn't have a technology-specific meaning. In his words, 'A.I. is the science and engineering of making intelligent machines.' The creator himself saw A.I. as not the ends, but the means. A.I. is not a technology any more so than physics or civil engineering. The challenge is, like all sciences, there must be application to bring concept to reality."

Walch, Managing Partner & Principal Analyst at A.I. Focused Research and Advisory firm Cognilytica, continues, "It might be more helpful to think of A.I. as a goal. If A.I. is considered to be a collection of technologies, then you can argue all day about what is and what isn't A.I. Are software robots A.I.? Are self-driving cars A.I.? Is computer vision A.I.? Is character recognition A.I.? If you think about it as technology then it's always subject to disagreement and interpretation. However, if you think about it as a goal, or a quest, then it's something we're always striving to achieve, even if we aren't quite there yet."

She concludes. "Forward thinking organizations and individuals understand that A.I. is not a technology. They don't ask whether something they implemented is or isn't A.I. Instead they ask what transformative effect that technology has. They ask what benefit they can realize from machines that can think and act as humans do. They ask how A.I. will represent opportunities to dramatically increase efficiency, reduce expenses, increase customer satisfaction, improve existing products and services, and create new business opportunities. Because at the end of the day, an organization itself is not about its technology, but its overall mission and objective. And just like those organizations, A.I. is not defined by technology, but by the overall objective."

Source: *Forbes, "Artificial Intelligence is Not a Technology," by Kathleen Walch, May 1, 2018, https://www.forbes.com/sites/cognitiveworld/2018/11/01/artificial-intelligence-is-not-a-technology/?sh=7c24bbf55dcb*

Steve Bennett, Director of Design Technology for 110-person architecture, strategic planning and data analytics firm Taylor Design, says, "In the early 2000s in Silicon Valley, everything had to have an "i" in front of it. Now it's an " A.I." The companies have had the last year to put a marketing spin on A.I. and put it on everything. Some people are going to fall for it, but we have to take a step back and ask, 'is this really A.I.?' It matters and there are ramifications whether something is or isn't A.I."

Chatbots are a good example. In its paper, "What is a Chatbot," IBM says, "The terms chatbot, A.I. chatbot and virtual agent are often used interchangeably, which can cause confusion. While the technologies these terms refer to are closely related, subtle distinctions yield important differences in their respective capabilities."

The paper continues, "Chatbot is the most inclusive, catch-all term. Any software simulating human conversation, whether powered by traditional, rigid decision tree-style menu navigation or cutting edge conversational

A.I., is a chatbot. Chatbots can be found across nearly any communication channel, from phone trees to social media to specific apps and websites."

Bennett says, "When I think of A.I., I think of a purpose-built A.I. system like Midjourney, which helps you create imaginary imagery. Or ChatGPT, which helps you create text-based imaginary things. Adobe Photoshop has an A.I. plug-in you can use to embed images with text-based prompts. Some people call their automated chat systems 'A.I.' when it's not. You can actually chat with an A.I. chatbot, but a non-A.I. chatbot is usually the one that you don't get an answer from, so you go around it and talk with a human."

A.I. STRENGTHS & WEAKNESSES

If there's so much confusion about what A.I. is, does it really matter whether a machine actually exhibits artificial intelligence if it does the job it's asked to do? In many cases, yes. A process or program functioning at the A.I. level holds many advantages over less sophisticated options.

Here's a look at some of the strengths and weaknesses of A.I. as it exists today. Strengths tend to focus on characteristics of efficiency, speed and precision. Weaknesses are more diverse, ranging from ethics to errors to security.

A.I.'S STRENGTHS

- **Data Processing and Analysis.** A.I.-equipped tools excel at processing and analyzing vast amounts of data quickly and efficiently, enabling insights and decision-making that would take far longer and be far more difficult for humans. Generative A.I. design tools can explore numerous design possibilities based on specified parameters far faster and more extensively than the industry previously could. A.I. can conduct performance analysis to analyze environmental and structural data to optimize building performance, or simulate and assess the energy efficiency of buildings.

- **Predictive Analytics.** "As an industry, from a reporting and analytics perspective, we're kind of reactive," says Lucas Hayden, Sr. Director of AEC Strategy for software solutions company Unanet. "We look at what has happened. With the predictive analytics you get with A.I., we look at what will happen. We can determine the best clients and projects to pursue, or when we're going to need to hire more people for a particular set of skills for work in the pipeline. There's so much opportunity to move to a more predictive and prescriptive way of running our projects and business."

- **Automation.** A.I. automates repetitive and routine tasks, leading to increased productivity and efficiency. This became a major factor for AEC firms worldwide in 2023. From cover letters to social media posts to expanding design options, architecture and engineering firms began automating many of their most mundane and time-consuming tasks. This is critical for an industry that has failed to improve its productivity despite the many productivity-improvement tools it now uses. (More on this later.)

- **Pattern Recognition.** Machine learning algorithms are adept at identifying patterns and trends within data, contributing to better predictions and problem-solving. LLMs and GPT recognize and generate human-like language patterns, while some neural networks perform non-textual pattern recognition tasks.

- **Precision.** A.I. systems can perform tasks with greater precision and accuracy, reducing the margin of error when executing specific functions. For instance, A.I.-powered robotic systems can lay bricks, pour concrete, or perform other tasks with a level of accuracy that may be challenging for human workers alone.

- **Personalization.** A.I. enables more efficient personalized experiences in various business areas, such as marketing and business development campaigns, customer relationship management (CRM) programs, content curation and design preferences. This also extends to features like smart homes to create personalized and adaptive environments. A.I.-infused systems can learn user preferences for lighting, temperature, and other factors to provide a customized living experience.

- **Learning and Adaptation.** Machine learning algorithms can learn from new data, adapting to changing circumstances and improving their performance over time. A.I. tools can learn from user feedback and design iterations, enabling architects and engineers to refine and optimize their designs based on real-world usage and performance data.

A.I.'S WEAKNESSES

- **Lack of Understanding and Common Sense.** A.I. systems lack a deep understanding of context and may struggle with tasks that require commonsense reasoning or a broader understanding of the world. This is why A.I.-generated content is usually easy to identify, and also why human involvement and curation of A.I. content is critical.

- **Propensity for Mistakes.** The performance of A.I. systems depends heavily on the quality and representativeness of the data it accesses to train itself. Biased or incomplete data can lead to inaccurate results. Like most human "know-it-alls," generative A.I. is "often wrong, but never in doubt."

- **Limited Creativity and Intuition.** A.I. struggles with tasks that involve true creativity, intuition, and an understanding of abstract concepts. This is one reason why direct-to-customer A.I. content is a recipe for disaster, especially in creative professions like architecture and engineering.

- **Bias and Fairness Issues.** A.I. algorithms will often mimic the biases that it encounters in training data, leading to unfair or discriminatory outcomes. Ensuring fairness in A.I. systems is a significant challenge.

- **Ethical Concerns.** A.I. raises ethical questions, especially in areas like privacy, accountability, originality and authenticity. People need to be careful using A.I. in sensitive applications.

- **Intellectual Property Uncertainty.** This involves at least two separate matters: 1) who owns Gen A.I.'s output; and 2) which and how much information should you share with it.

- **Zero Emotional Intelligence.** The current state of A.I. lacks emotional intelligence and cannot understand or respond to human emotions. Though the day may come where this is not the case, at its core, A.I. functions as a machine like any other machine.

- **Lack of Regulation and Constraint.** When ChatGPT exploded on the scene, the use of generative A.I. products rapidly spread throughout industries and companies, often with little or no oversight. Smart leaders and companies have reined in the use of A.I. with policies and best practices, but there are still – and will continue to be – rogue and dangerous use of A.I. programs, especially as technology keeps marching fast forward. Deepfake images meant to mislead, misinform or embarrass someone are an example.

- **Security Risks.** Similarly, there is so much unknown about what A.I. does and can do. A.I. systems may be vulnerable to adversarial attacks. For example, malicious actors could manipulate or input deceptive data, leading to incorrect and potentially dangerous outputs.

- **High Resource Requirements.** Some A.I. models, especially deep learning models, require substantial computational power and energy resources, posing environmental, financial and social concerns.

Understanding the strengths and weaknesses of A.I. is essential for responsible development, deployment, and use of A.I. technologies. Addressing ethical considerations, bias mitigation, and ensuring transparency are other critical steps in maximizing the positive impact of A.I. while minimizing potential drawbacks.

SIDE NOTES: THE TURING TEST

Machine learning pioneer Alan Turing created an A.I. litmus test in 1950, basing it on "The Imitation Game" in which a judge tries to determine which of two participants is the woman and which is the male. Turing changed the game to include a machine participant, a human participant and a human judge, with the judge's task to determine which of the participants is human and which is the machine. If the judge is wrong at least half the time, The Turing Test deems the machine to be exhibiting A.I.

Advances in A.I. technology challenge The Turing Test's reliability, and other methods have been developed. In an article on the topic, technology marketing company TechTarget describes some of the several variations to the Turing Test:

- Reverse Turing Test – where a human tries to convince a computer that he or she is not a computer. An example of this is a CAPTCHA.

- Total Turing Test – where the questioner can also test perceptual abilities as well as the ability to manipulate objects.

- Minimum Intelligent Signal Test – where only true/false and yes/no questions are given.

Alternatives to Turing Tests were later developed because many see the Turing Test to be flawed. These alternatives include:

- The Marcus Test – in which a program that can 'watch' a television show is tested by being asked meaningful questions about the show's content.

- The Lovelace Test 2.0 – a test made to detect A.I. through examining its ability to create art.

- Winograd Schema Challenge – a test that asks multiple-choice questions in a specific format.

"Although the variations of the Turing Test are often more applicable to our current understanding of A.I., the original format of the test is still used to this day. For example, the Loebner Prize has been awarded annually since 1990 to the most human-like computer program as voted by a panel of judges. The competition follows the standard rules of the Turing Test," writes TechTarget.

"Despite much debate about the relevance of the Turing Test today and the validity of the competitions that are based around it, the test still stands as a philosophical starting point for discussing and researching A.I. As we continue to make advances in A.I. and better understand and map how the human brain functions, the Turing Test remains foundational for defining intelligence and is a baseline for the debate about what we should expect from technologies for them to be considered thinking machines," TechTarget writes.

https://www.techtarget.com/searchenterpriseai/definition/Turing-test

A.I. IN EVERYDAY LIFE

A.I. has become increasingly integrated into various aspects of everyday life, impacting both personal and professional spheres. Here are some common areas where you can find A.I. applications:

Virtual Assistants. Virtual personal assistants like Siri, Google Assistant, and Alexa use A.I. to understand and respond to voice commands, manage calendars, provide weather updates, and more.

Smart Home Devices. Smart home systems use A.I. to learn how to better automate tasks such as adjusting thermostats, controlling lighting, and managing security cameras based on user preferences and behavior.

Recommendation Systems. Online platforms, such as streaming services (Netflix, Spotify), e-commerce websites (Amazon, eBay), and social media (Facebook, Instagram), use A.I. algorithms to provide personalized recommendations based on user preferences and behavior.

Search Engines. Search engines like Google use A.I. to enhance search results, improve relevance, and predict user intent, making it easier for people to find information quickly.

Healthcare. A.I. is employed in medical imaging for diagnosis, drug discovery, personalized medicine, and predictive analytics. Smart chatbots also provide initial patient interaction and assistance.

Finance. A.I. is used in fraud detection, algorithmic trading, credit scoring, and customer service in the financial industry.

Autonomous Vehicles. A.I. plays a crucial role in self-driving cars, enabling them to perceive their environment, make real-time decisions, and navigate safely.

Language Translation. A.I.-powered translation services like Google Translate use machine learning to provide accurate translations between languages.

Email Filtering and Spam Detection. A.I. algorithms help filter emails, identify spam, and categorize messages in email platforms.

Gaming. A.I. creates non-player characters (NPCs) with realistic behavior and adapts the game environment based on player actions.

Social Media. Social media networks use A.I. for content moderation, image recognition, and personalized content delivery on platforms like Facebook, Twitter, and Instagram. While it can be a powerful and positive tool, the use of A.I. is also causing controversy with complaints of social media companies abusing it for commercial, political and ideological reasons.

Customer Service. Customer support services use chatbots and virtual assistants to provide instant responses to common queries and assist with issue resolution.

Education. A.I. applications in education include personalized learning platforms, intelligent tutoring systems, and automated grading systems.

Manufacturing and Logistics. A.I. is used in predictive maintenance, quality control, and supply chain optimization to improve efficiency and reduce costs in manufacturing and logistics. This type of us offers substantial potential to improve the project delivery and project management process in the AEC industry.

Weather Forecasting. A.I. models analyze large datasets to enhance the accuracy of weather predictions and improve forecasting capabilities.

A.I.'s involvement in various aspects of modern life sometimes goes unnoticed by the typical user, but the convenience, efficiency and direction that it often provides has become indispensable. As technology progresses, the integration of A.I. into everyday life will only increase.

OTHER TECHNOLOGIES THAT MAY OR MAY NOT USE A.I.

Here are 10 current technologies popular in the AEC industry that may or may not incorporate A.I. In many cases, such as with AR/VR and digital twins, the introduction of A.I.-driven tools and components can vastly improve the capabilities and value of the technology, but they are not essential to their basic function (i.e., they have a use even without A.I. as a factor).

- Building Information Modeling (BIM)
- 3D Printing
- Augmented Reality (AR)
- Virtual Reality (VR)
- Drones
- Quantum Computing

- Digital Twins
- Construction Robots
- LiDAR Technology
- Simulation Software

WHAT'S NEXT FOR A.I.

Predicting the future of A.I. requires considering both the potential advancements and the challenges that may arise. While it is impossible to provide a definitive roadmap, here are some trends and possibilities that could shape the future of A.I.:

Advancements in Deep Learning
Deep learning, a subset of machine learning, has shown remarkable success in various applications. Future advancements may lead to more sophisticated neural network architectures and training techniques, enabling A.I. systems to handle increasingly complex tasks.

Explainable A.I. (XAI)
Addressing the black box nature of some A.I. models, there is a growing emphasis on developing explainable A.I. Future A.I. systems may be designed to provide transparent explanations of their decision-making processes, fostering trust and understanding.

A.I. for Good
The use of A.I. for solving societal challenges, such as healthcare, climate

change and poverty, is likely to increase. A.I. could play a crucial role in addressing global issues and improving the quality of life for people around the world.

A.I. in Healthcare

A.I. applications in healthcare, including medical diagnosis, personalized medicine, and drug discovery, will likely expand. A.I. may contribute to more accurate diagnoses, improved treatment plans and better patient outcomes. Biotech company Insitro is using A.I. to help treat diseases such as ALS, cancer and liver disease.

Edge A.I.

Edge computing, where A.I. processing occurs on devices rather than relying solely on centralized cloud servers, is becoming more prominent. This approach can lead to faster response times, increased privacy, and reduced dependence on constant internet connectivity.

A.I. Ethics and Regulation

As A.I. adoption increases, there will likely be a greater focus on ethical considerations and regulations. Efforts to ensure fairness, accountability, transparency and responsible use will shape the development and deployment of A.I.-driven technologies. Model Evaluation & Threat Research (METR) is a non-profit group that assesses if A.I. systems pose a threat. In 2023, they were working with tech companies Anthropic and OpenA.I. to evaluate their systems.

Human-A.I. Collaboration

Future A.I. systems may be designed to collaborate more seamlessly with humans, complementing human skills and decision-making rather than replacing them. This could lead to new forms of human-A.I. partnerships.

Autonomous Systems

Advances in autonomous systems, such as self-driving cars and drones, will continue. The development of robust A.I. algorithms for navigation and decision-making could lead to increased autonomy in various domains.

Natural Language Processing (NLP) Improvements

Natural language understanding and generation capabilities continues to improve, which will lead to more sophisticated chatbots, virtual assistants, and language translation services.

Quantum Computing Impact

The development of quantum computing could have a massive impact on A.I. Quantum algorithms may significantly speed up certain A.I. computations, enabling the solution of complex problems that are currently intractable. Much of the theoretical possibility of A.I. could be unlocked by quantum computing.

FUTURE CHALLENGES AND CONSIDERATIONS

As A.I. grows more sophisticated and is used more broadly throughout business and society, the challenges and concerns surrounding it will grow as well. Here are a few to watch:

Ethical Concerns

As noted earlier in this book, addressing ethical challenges, including bias, fairness, and privacy, will be crucial to ensure the responsible development and use of A.I.

Regulatory Landscape

The establishment and adaptation of A.I. regulations and standards will play a vital role in guiding its development and deployment. As calls for increased oversight and regulation of A.I. grow, the U.S. government moved the issue forward with its Blueprint for an A.I. Bill of Rights in October 2022, and the 53-page Executive Order on the Safe, Secure, and Trustworthy Development and Use of Artificial Intelligence in October 2023. The Biden Administration says that the latter "significantly advances the United States' policy framework regarding [A.I.]."

Job Displacement

The potential for job displacement due to automation raises concerns, and efforts to reskill the workforce will be essential. Few AEC firms and professionals seem aware of and prepared for the coming changes that the advanced use of A.I. will cause in a typical workforce.

Security Risks

As A.I. systems become more integrated into critical systems, the risk of security vulnerabilities and adversarial attacks is likely to increase.

A.I. and Creativity

The extent to which A.I. can contribute to creative tasks, such as art, music, and innovation, remains an open question with ethical and philosophical implications.

We address these and other threats further in Chapter 4: Best Practices for AEC Firms and Chapter 5: Major Obstacles & Concerns.

CHAPTER 1 TAKEAWAYS

Here are the key takeaways from this chapter:

- There is substantial confusion about what artificial intelligence (A.I.) actually is.
- A.I. is a process or a goal, rather than a technology.
- Machine Learning and Deep Learning are two major components that allow machines to achieve A.I.
- There are two major types of A.I. – Narrow or Weak A.I. and General or Strong A.I.
- All current A.I. is narrow; general A.I. (AGI) is theoretical at this point.
- Strengths tend to focus on characteristics of efficiency, speed and precision (e.g., Predictive Analytics, Automation).
- Weaknesses are more diverse, ranging from ethics to errors to security.
- A.I. is permeating nearly every aspect of everyday life, a trend that will only expand.
- Many technologies are lumped in with A.I., but may or may not represent true artificial intelligence. Examples include BIM and AR/VR.
- The future of A.I. appears limitless, but the challenges are daunting.

Chapter 02

CURRENT USES OF A.I. IN AEC

It is difficult to measure the extent and timing of the disruption that A.I. will cause in the AEC industry. Firms are wading in, often driven by a grass-roots approach in which staff independently identify and encourage the use of products they discover. The catalysts could be younger, lower-level employees who encounter a technology outside of work, or even C suite occupants.

VHB's Mulholland serves on the American Council of Engineering Companies (ACEC) Technology Committee, which was founded in 2023. He is a registered traffic engineer, but moved into the CTO role in 2021 "to manage and drive the firm's strategy for investing in, using, and developing the technologies that are increasingly important to the performance and success of AEC firms."

He adds, "I joined ACEC's technology committee to help better position VHB and the industry from technology-related shocks and disruptions that are facing us."

The prevailing opinion is that most AEC firm leaders are unprepared for the sea change that A.I. is likely to deliver in the coming years. They're intrigued by A.I., even dabbling in it, but they aren't yet able to fathom the extent to which A.I. is likely to change the industry.

The AIA/Deltek Architecture Billing Index (ABI) from September 2023 found that "while few firms of any size are currently using A.I. technology, more than two-thirds of large firms plan to implement A.I. for design-related applications in the next decade (with 45% planning to do so in the next two years), and 60% of large firms plan to implement A.I. for firm operations applications in the next decade (with 34% planning to do so in the next two years)."

K.P. Reddy is founder of Shadow Ventures, which invests in seed-stage construction technology and property technology companies. In a 2023 episode of PSMJ's «$10,000 Hour webcast,» he noted that A.I. has many AEC leaders worried, but that this lack of preparation is attracting venture capital firms who see the potential for it to improve productivity and increase value for the industry.

"I don't have to scare people that much; everybody's already kind of scared," he said. "Most of the mainstream VCs we work with are looking for old-line industries like ours because they have the maximum opportunity for disruption."

Reddy says this is why the current state of A.I. in the industry is unique compared with other recent technology developments. "What's really shifted is that A.I. is now in the hands of the consumer," he says. "I have the good fortune of talking to 20 CEOs a week in the AEC industry, and not since the Blackberry have I seen executives playing with a toy like the one called ChatGPT. It's not IT going to the CEO saying, 'Hey, let me show you this.' It's the CEOs saying, 'What are we doing about this? I'm playing with it with my kids, and this is fascinating.' This is the first time I've seen this type of a top-down push, and the executives are trying to reimagine what their firm may look like with it."

A.I. IN AEC: MYTH VS. REALITY

"There's a lot of marketing and a lot of news hype that's increasing confusion," says Frank Lazaro, A.I. Expert and Founder of AE Marketer. "Most people don't know or understand what is real and what is the art of the possible. There's a little bit of overestimation of A.I.'s impacts today, in terms of what it's actually doing for the industry versus what it has the potential to do for our industry."

Which raises the question: How are AEC firms using A.I. in 2024?

In his December 2023 PSMJ presentation, "A.I. in AEC 2.0: The Executive's Guide to Navigating Opportunities and Challenges," Lazaro said that A.I. use in AEC firms was almost exclusively occurring in sales and marketing as 2023 turned to 2024, but he expects to see increasingly common use of A.I. in back office and project delivery functions in the near future.

"Where A.I. is happening right now in our industry is around content generation," says Lazaro. "It's around social media posts, rewriting proposals and project sheets; it's all around the basics."

Lazaro continues, "The second phase is back office, which I think of as HR, accounting, those types of functions. Using A.I. to write your job descriptions, rewriting benefits pages, employee newsletters; doing things where you want to create new content for more of those internal or back office areas."

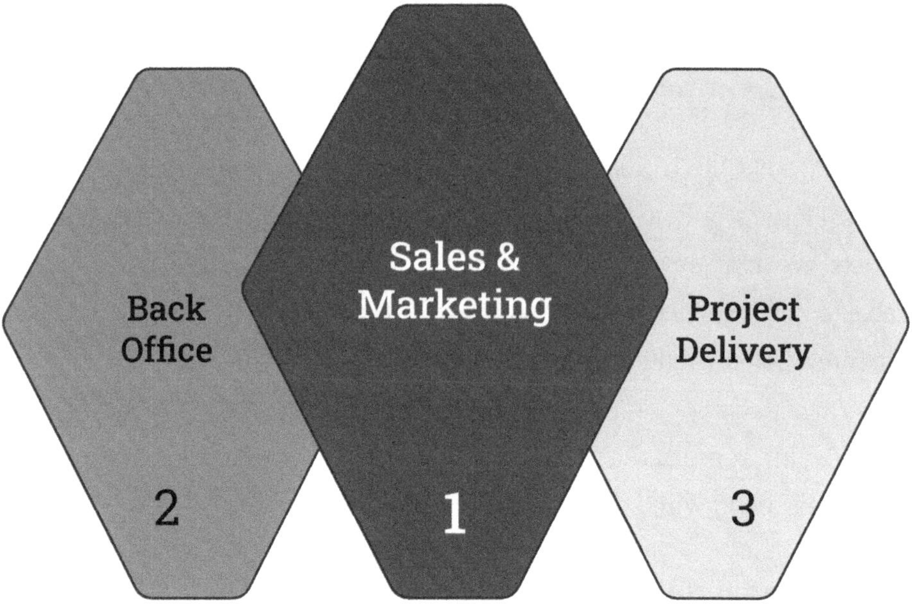

Frank Lazaro's Three Phases of Current A.I. Use in AEC

The last phase to broadly incorporate A.I. is project delivery, says Lazaro. "The way project delivery is going to be impacted is that the tools that you use on a day-to-day basis are eventually going to incorporate A.I. The Autodesk website has a whole section on how they're incorporating A.I. into their various tools."

Unanet's Hayden says, "Speaking generally, there's a lot of technology already in the hands of our users. A couple of specific examples are predictive modeling that's forecasting revenue for projects. Financial automation is also a key one, where firms are using A.I. technology to process expense receipts and invoices. Another example is contact detection that uses machine learning to find contact details in your email and updates the related records – without leaving Outlook. This is in Unanet products as well as others."

A.I. USE IN SALES & MARKETING

Justin Hoff, COO of 14-person Iowa-based Martin Gardner Architecture, P.C. (MGA) says, "It's a great benefit to smaller firms to use ChatGPT to write proposal letters, or help with smaller social media and graphic design kinds of posts. We don't have a creative writer or graphic designer on staff, so A.I. is able to help us. We use Adobe for some generative A.I. imaging. It helps us generate a better product to the client in a tighter time frame."

Following are some of the components of sales and marketing where A.I. is being used, along with some examples from practitioners.

CONTENT MARKETING

In an earlier video presentation for PSMJ, Lazaro said he could do a LinkedIn post in under five minutes by feeding a question into an A.I. program – he used the example of "technology's impact on the AEC industry." He said he does this approximately once a week, receiving seven paragraphs of written content with hashtags. He reads it, edits it for tone and voice, and then posts it on LinkedIn.

Appearing on PSMJ's November 2023 "$10,000 Hour" webcast, Scott Steiding, president of consulting firm Productive Pursuits, referenced a firm whose creative writer left unexpectedly. Rather than replace him, the firm called on a member of its marketing team to use ChatGPT to create thought leadership content. "The ability to go quickly from ideation or a discussion with a subject matter expert to a first draft is crazy fast," said Steiding.

Here are three ways that content writers can use generative A.I. programs – in this case, ChatGPT – to their benefit, without wholly giving over responsibility to the machine.

1. **Simplify your search.** Use ChatGPT as an enhanced search engine. While Google and its many rivals deliver sources of information in list form that you can scan until you find one to pursue further, ChatGPT instantly consolidates and organizes data into article format. A content creator can review the article it produces, determine which parts of the content deserve further research or discussion, and build on it from there.

2. **Brainstorm.** Every writer suffers now and then from "blank page syndrome." When faced with a writing assignment, but unable to find the first words, ask a relevant question of ChatGPT and it will usually spit out content that can trigger creativity. "It's a lot easier to start a report with a halfway filled out piece of paper," Bartolf says. "Let A.I. go crazy, then look at it and change this, this and this. Add the human element to it. Use A.I. to automate the beginning processes, the labor intensive processes."

   "I've used Midjourney to develop illustrations for a presentation," says Taylor Design's Bennett. "I ask it how it would illustrate a particular part of the presentation for people, and it takes it and runs with it. Then I'll massage it. It gets me going. Gives me ideas. Sometimes I ask ChatGPT a kind of stupid question, and occasionally it will surprise me with a good answer."

3. **Find the Missing Piece.** Generative A.I. is a blessing for the absent-minded. After a content creator completes an article, it's worth checking with ChatGPT for anything overlooked. Often, the program will find something you missed.

Content creators need to be careful, however. Most tech-savvy leaders know that serving up the results directly from a generative A.I. program is unwise and unsafe. Here are three reasons why:

1. **Mind the source.** The open version of ChatGPT just grabs what it can from the Internet and spits it back out. Most of the content is generic, reducing the threat of copyright infringement or other legal concerns. However, the content usually lacks attribution. Newer programs, including Microsoft's free version of Copilot, include sources of the data, easing the concern. Either way, be extremely wary of publishing anything to a wide, public audience or directly to clients without verifying the source.

2. **Recognize its fallibility.** Generative A.I. programs are supremely confident about their output, but are sometimes wildly off base. When distributing anything derived from a program like ChatGPT, you must confirm that it's true.

3. **Add humanity.** Content that is 100% generated by the A.I. programs available in early 2024 shared a tendency to be dry and expectedly robotic. Many – if not most – readers can sense that something is lacking. For this reason, as tempting as it might be as a timesaving exercise, don't just cut and paste the output from an A.I. program. Edit it and add the human touch. Otherwise, it's not only lazy and bordering on dishonest, the content you push out may reflect negatively on you and your organization.

The internet is full of advice on how to use A.I. for content writing, and what to avoid. For instance, on the newsletter website medium.com, James Presbitero Jr., published the article "These Words Make It Obvious Your Text Was Written By A.I."

Presbitero writes, "We all know what "human writing" sounds like. But until recently, I still couldn't put a finger on it. Last November, I got a job as a content writer and editor. Since then, I've written and edited 12 to 20 articles every week. It was through this repetitive, highly systematic, and quite stimulating job responsibility that I noticed consistent giveaways to A.I. output."

Presbitero's red flags are the words and phrases:

1. Transformative
2. Foster/fostering
3. Tapestry/A tapestry of...
4. This is about ... / All about ...
5. Think of X as ...
6. Like/It's like ...
7. Not only ... but also ...

Presbitero adds that another giveaway is sentence length that doesn't vary the way most good writing does.

Not everyone agrees that A.I.-written content is easy to spot. Toby McInnis, in his article "There Are No 'Special Words' That Give Away A.I.-Generated Content," asserts that "A.I. writing has the same faults as human writing."

McInnis' take is cynical, arguing that a machine trained on bad writing will give back similarly bad writing. He argues:

- Bad writing is easier to produce than good writing, therefore more bad writing is produced.

- That means A.I. is trained on more bad writing than good writing.

- And that means A.I. (rightly) predicts that the next word the average human would choose in any given sequence will be a dumpster fire decision guaranteed to make any half-sentient being either giggle or retch.

McInnis' article veers suddenly toward a far more hopeful tone. He writes:

> *So here's my hot take: the fact people are so eager to claim content is written by A.I. is a valuable signal. It shows that the risk moving forward is not that we will be constantly tricked by devious A.I.-generated content, but that we will lose trust that any message is the product of a real, thinking and feeling human.*
>
> *This presents an opportunity for business content. It will dramatically increase the value of factors that are still the exclusive domain of human content creators. These include:*
>
> - *Genuine excellence in prose and design*
> - *Humor and irrational ideas*
> - *Utterly unnecessary levels of precision and attention to detail*
> - *And thought leadership born of actual first-hand experience*
>
> *I also believe this is the most valuable use-case for A.I. in commercial content production: arriving at the "differentiation" part of the process faster.*
>
> *ChatGPT can show you all the most obvious, cliched solutions to the problem you're solving; serve up all the most generic topics, insights and advice; and*

enable faster market research to figure out how you can ensure you don't regurgitate the same old pap.

It can aggregate all those mindless "trends" articles you spend the festive season yawning over, and show you what the average person in your industry thinks – so you can actually make a reasonable attempt at "leading" them somewhere more interesting.

And it can remind writers that their most valuable skill isn't producing lots of words, or meeting that deadline – it's reading, appreciating and obsessively pursuing real quality content.

Presbitero and McInnis came at the issue from different directions, but essentially wound up in the same place. Whether it is repetitious buzz-terms or just plain bad writing, A.I.-generated content has significant limitations that leave it incapable of fully devaluing human writing.

Yet, generative A.I.'s use in content marketing is certain to continue growing in AEC firms, particularly as we become more comfortable with the tools and the tools become even more sophisticated. Done correctly, marketing content is among the lowest-risk uses, while offering massive benefits in efficiency and value.

In an article on its website called "How to Put A.I. to Work for Better Content Marketing," the Content Marketing Institute (CMI) suggests sharpening the audience's understanding of the process. "Create an A.I.-powered marketing persona," says CMI.

Its sample queries include:

- What does your audience worry about?
- Where do they get their information?
- What topics are most helpful for their jobs to be done?
- What triggers them to go looking for answers or help?
- What information do they need before they're willing to become a lead?

The article then quotes Andy Crestodina, co-founder and CMO of Orbit Media Studios: "Unlike humans, you can quiz your new A.I. persona day or night. It never gets tired of sharing insights with you."

PROPOSALS & PRESENTATIONS

Proposal creation using A.I. offers many of the same benefits and warrants some of the same concerns as content marketing.

Specific to proposals, marketing teams are tapping generative A.I. to develop cover letter, interesting imagery and boilerplate content. Some are using it to show insight into design concepts. As MGA's Hoff said, A.I. tools offer great benefit to small firms with limited marketing staff to work on proposals.

Products that aid in these areas constitute a growth industry. For example, techopedia.com lists the following products as the top five cover letter writing programs.

- CopyA.I. – Overall, the best cover letter writer, with brand voice options and 95+ languages

- Rytr – The best free cover letter writer with 40+ use cases, 20+ tones, and A.I. image generator

- Simplified – One of the most easy-to-use A.I.s with one-line prompts and minimal grammatical errors

- Jasper A.I. – Popular A.I. writer with a handy Content Improver tool for grammatical accuracy and A.I.DA template

- Word A.I. – Tailor-made for those who are afraid of having their A.I.-generated cover letters detected

In addition to content and image creation, here are some ways that A.I. can (or will) enhance the proposal process and presentations for AEC firms:

Generative Design. The proposal team can use A.I. to expand the range of proposed design options based on specified criteria. By relying on the speed and efficiency of an A.I. program, the team can create and assess more designs, and offer more options to the client.

Parametric Design Tools. Similarly, the proposal team can create designs that dynamically respond to specific parameters solicited from the RFP or the project owner.

Proposal Review & Improvement. A.I. can analyze and optimize proposal content based on historical data, successful (and even failed) past proposals, and industry standards. One substantial benefit is that firms can target specific weaknesses identified internally or externally while conducting their review to ensure against repeat errors.

Virtual and Augmented Reality (VR/AR). There is a gray area surrounding the question of whether VR/AR is technically A.I. or just an example of 3D technology. While researchers at Stanford are using A.I. to improve 3D displays for virtual and augmented reality technologies, the typical use of VR/AR in a presentation or interactive proposal would not strictly constitute A.I. Nonetheless, this is an evolving technology with endless potential.

Project Planning and Scheduling. A.I. algorithms can analyze historical project data for risks, obstacles, milestones and so on, to create more accurate and realistic design and construction schedules.

Cost Estimation. Similarly, A.I.-driven tools can analyze historical project data, market conditions, and material costs to develop better cost estimates.

Editing. Natural Language Processing (NLP) tools can improve the clarity and coherence of proposal content, seeking out and eliminating jargon, repetition, misused words and so on. One product example is Wordtune, which helps rework sentences by offering word alternatives.

Risk Assessment. Clients hate risk. An A.I.-powered risk assessment drawing on prior project data (yours and others) that highlights risks and presents strategies for mitigating could tip the scales in your favor.

Visual Recognition for Documentation. Poor data organization is a cross that many AEC firms bear, proposals and presentations included. A.I.-driven visual recognition tools can search for and locate relevant past proposals and presentations to streamline the organization and creation process.

Collaborative Tools. Collaborative software tools have been used for many years by marketing and proposal development teams, but the widespread availability of generative A.I. products enables faster and more efficient collaboration. Products already on the market promote more efficient workflows, help to generate ideas, and automate certain processes.

Integration with Project Management Systems. In some enterprise systems, A.I. can help integrate the proposal and project management processes.

GO/NO-GO DECISIONS

It is difficult to find many firms today that are using predictive analytics to gauge the likelihood of winning or making money on a project based solely on the contents of a request for proposal (RFP). But some apparently are.

Lud Eng, Director of Information Technology for Stratus, says he has seen a bit of this in action. "On the text-based side, I've seen people have a lot of success using OpenAI and some of the other public tools. Firms can get a really quick idea of whether they want to spend time doing a proposal."

As predictive analytics in A.I. improve, and AEC firms grow more comfortable with the technology, the use of A.I. in the go/no-go process will grow exponentially.

IMAGE GENERATION

Midjourney is one of the most popular products using A.I. for image creation.

"To be worthwhile, A.I. has to be transformative in your workflow," says Taylor Design's Bennett. "Midjourney helps create imagery that otherwise a designer would have had to flip through a magazine or watch a video to come up with ideas. A.I. puts it in the computer and the designer synthesizes it in their brain and creates new design ideas for projects. It automates something in a workflow that traditionally would have been human and manual."

"Adobe has some new generative tools that are in a suite that people are already familiar with," says Eng. "There are no intellectual property issues; it's licensed to users already. That's an easy, low-risk point of entry. A lot of its use is for graphics in marketing materials and social media. It can make reports, even internal ones, look more professional by customizing the graphics to the report."

The use of A.I. in imagery is particularly controversial, and thus potentially fraught with danger. The ability to create realistic-looking images, then represent them as real, is of massive concern far beyond just the AEC space. Stories abound about the threat that A.I.-generated imagery presents, and about how quickly the role of photographers will change or even become obsolete.

Thomas Smith publishes a newsletter on medium.com called The Generator. In a January 2024 article titled "Midjourney 6 Means the End for a Big Chunk of the Photo Industry," he writes about a conversation he listened in on at a digital media conference in 2019:

"Chatting casually among themselves, these leading A.I. researchers and practitioners agreed on one thing: image tagging is cool, but give the tech a couple of years, and it will be generating new images based on a couple of sentences of text. At the time, this seemed ludicrous. A.I. image generation was in its early, early infancy. It would still be two years before OpenA.I. would release the first version of DALL-E. And even when that happened, the initial output of A.I. image generators wasn't exactly impressive."

Smith continues, "In just a few short years, A.I. image generators surged ahead and were generating images that rivaled real photographs in their detail, realism, and visual power. We went from grainy, low-res, clipart-like images to beautiful, photorealistic [renderings] in just under three years. It's mind-blowing."

As Smith notes in his article, the output from these image-generating tools has progressed from cartoonish to scarily realistic. "This presents a huge problem for people who make a living creating simple, aspirational photos of people," writes Smith. "Given a choice between an expensive and labor-intensive photoshoot and A.I., most customers seeking a stock image will choose the latter option. Before, Midjourney's subtle errors made the system's human-focused images essentially unusable. Now, they may not be perfect, but they're good enough. And in a commercial space, 'good enough' always wins."

What does this mean for photography in AEC? Almost certainly it means that money once spent on professional photographers will shift toward computer programs. Though it may be difficult to reproduce the perfect image of a completed project with an iPhone photo or a few lines of descriptive text, technology already allows for the manipulation of images to eventually land where the user wants it to land. Even if the representation of a subject isn't true-to-life, that may not matter – close enough might not only be good enough; it might even be better.

While the more basic images used for marketing, operations and project-related purposes will likely go the way of the computer, some of the graphical representations that the AEC industry produces are unique enough to require the hand of a human photographer or graphic artist for the foreseeable future.

CUSTOMER RELATIONSHIP MANAGEMENT (CRM)

The functional automation found in A.I.-focused technology is a natural fit for the often-tedious and mistake-riddled process of creating, managing, and enhancing a company's CRM. "You can use machine learning to take an email from Outlook, parse out the signature and add it directly into the system's record for the individual," says Unanet's Hayden. "It can go through email in a client- or project-aware posture, subject to the data that's held in the CRM. Contact enrichment. These are examples of how A.I. is being used today."

DDI Development, in an article titled "A.I. in CRM: Benefits and Use Cases of A.I.-Powered CRM System," says that NLP-driven CRM personalization allows companies to more efficiently group customers based on their behavior, preferences, and purchase history. "This can result in A.I.-enhanced customer journey mapping. With A.I.-driven customer interactions, customer experiences and data can be tailored to offer personalized recommendations and content that meet their individual preferences."

The impact of integrating artificial intelligence and automation with CRM by 2025

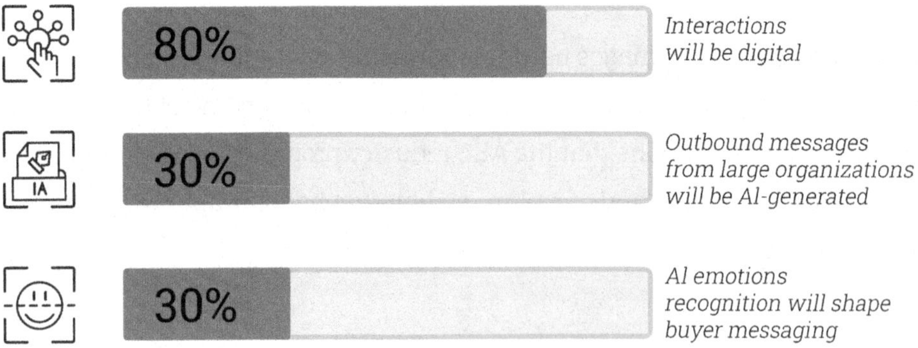

Source: Bitrix24

A.I. USE IN THE BACK OFFICE

As of early 2024, AEC firms were just beginning to recognize the benefits of A.I. in the human resources, accounting, and legal functions. In this, however, our industry is not alone. This is particularly relevant because so much of our industry is comprised of smaller firms with fewer resources to vet A.I. systems, and perhaps less incentive to take the leap due to uncertain return on the investment of time and money.

Yet, A.I. is coming quickly to the back office. "On the business side of the house, there's a lot of opportunity to harness predictive analytics and prescriptive insights" says Hayden. "For a long time, people have used earned value management and estimate-to-complete (ETC) to make sure they're running their projects profitably. Imagine a world where you could see very readily how your project is on track to perform. It could tell you things like, 'You have some issues in the back 30% of the project and here are some things you can do today to avoid those challenges altogether.' A.I. will be able to provide a vehicle for being able to forecast project performance and provide a means for prescriptive action in light of those elements."

Here are some ways that companies are using A.I. for back office functions:

HUMAN RESOURCES

Here are six areas where A.I. is already benefiting the HR function of companies inside and outside the AEC industry.

HR Administration. A.I. can help HR professionals automate repetitive and time-consuming tasks such as data entry, payroll, benefits administration, employee handbooks, job offers and so on. Some companies employ A.I. chatbots that provide around-the-clock HR assistance. Employees can chat about leave policies, benefits and all other areas of HR, within the limitations of the program, of course. This can increase the efficiency of the process and the HR team by automating these interactions.

Recruiting. A.I. tools can now screen resumes, schedule interviews using chatbots, and generate interview questions and personality assessments.

Onboarding/Offboarding. In an article called "7 Ways A.I. is Used in Human Resources," recruitment software company Workable writes about an A.I.-driven onboarding platform that walks a nervous new employee through the necessary forms, introduces her to company policies, and even sets up her initial training sessions. "Similarly, when an employee leaves the company, the A.I. can manage exit surveys, return of company property and other offboarding tasks, ensuring a smooth transition for both parties," the article states.

Employee Satisfaction. GenA.I. can help create and administer employee satisfaction surveys, offer personalized career guidance, develop and manage professional development planning and progress, and encourage employee engagement.

Training. "I use ChatGPT to take transcripts from videos and convert them into questions for training, to see if people are retaining what they're learning," says Bennett. "I'll take the transcripts from the videos and ask ChatGPT to create questions with four possible answers. I'll give it a prompt like, 'Summarize two main points from this text and give me two multiple-choice questions with four possible answers, then highlight the correct one.' ChatGPT is really good at guessing what's not right."

Workforce Planning. By analyzing data such as current workforce skills, industry trends, and company growth plans, an A.I. tool can predict how the organization chart might need to look in the coming months or years. This allows for more-forward-looking workforce planning.

Engineering and construction companies can leverage generative AI applications for value generation in 2024

Value realization	Use case	Application
Project management proficiency	Generation of optimized site plans	Develop site plans by automating certain aspects of the designn process
	Project scheduling	Optimize schedule by analyzing resource availability and various other constraints
	Content summarizatin	Extract key information from technical documents and synthesize field reports
Design and engineering adeptness	Efficient building automation systems	Recommend the design of efficient and sustainable building automation systems
	Generative design	Create new 3D product design based on inputs and constraints to visualize multiple aternatives of a particular industrial component
	Generative materials design	Utilize scientific principles to create new materials with optimized properties
Operational efficiency	Process optimization	Organize/reorganize project activities based on status and notify or prepare a contingency plan
	Smart summaries for drone surveys	Summarize large volume of drone footage and enable querying for various applications
Enhanced safety, risk management, and compliance	Quality assurance	Generate synthetic images required to train machines and generate preventive algorithms with automating site and quality inspection processes.
	Ensuring compliance	Automate compliance cheching of building designs against local building codes and regulations

Source: Deloitte AI Institute.

delotte.com/us/en,/ingmnts reseanch-centers/center-energy-industnals. html

ACCOUNTING

On the New Jersey Society of Certified Public Accountants (NJCPA) website, Sean P. Breheney of PKF O'Connor Davies writes in a September 2023 article ("The A.I. Revolution: Transforming Accountants' Roles"): "Despite having an historical reputation as a relatively conservative profession, accounting has been steadily shifting to embrace technological advancements, the most relevant of which is artificial intelligence (A.I.). Specific to the accounting industry, A.I. is used to automate repetitive, high-volume and time-consuming tasks traditionally performed by humans."

These tasks include data entry, reconciliation, classification and analysis of financial transactions. "Functions such as tax preparation automation and financial statement audit testing have also become targets of A.I. implementor," Breheney writes. "If implemented the right way, A.I. has the potential to significantly improve efficiency and reduce costs for the accounting industry while freeing up professionals' time to focus on higher-level, value-added engagements."

As it pertains to in-house finance and accounting departments for AEC firms, A.I. systems can help with customer interactions, decreasing busy work and reducing mistakes. A.I. can also help with regulatory compliance, auditing and reporting.

LEGAL

Though less commonly needed in an industry with as many small and midsized firms as AEC has, the legal component of doing business can also benefit from A.I. Here are a few ways:

- Automating routine and repetitive tasks such as data entry, contracts, document reviews and discovery.

- Analyzing large volumes of data, case law, precedents and regulations.
- Generating insight and recommendations based on previous legal strategies, scenarios and narratives.
- Monitoring legal compliance, obligations, risks and ethics.

A.I. USE IN PROJECT DELIVERY

The third leg of Lazaro's A.I. in AEC stool is the one likely to be slowest to achieve full fruition. But when it comes, it will come quickly and change the industry.

KP Reddy says, "My thesis is that if you're only going to use A.I. for the back office, the work the customer doesn't see, it's going to be 100% a race to the bottom.. If you move A.I. from the back of the house to front of the house, and start developing new products and services that engage in a new type of client experience, all of a sudden you're going to have a lot more value creation."

DESIGN & PLANNING

"There are great A.I. tools being used for layout planning right now," says Forestay's Bartolf. "You throw in 12 different variables – light, space, proximity of spaces to one another – feed it into A.I. and it can generate 30 or 40 layouts. And then you take the three most optimal ones."

"We use Dynamo Scripts in Revit, Midjourney for image preferences, and ChatGPT," says MGA's Hoff. "A lot of junior staff would hear about a new tool once a week, it seemed. They wanted to know what they could use, how to use it. We pumped the brakes and vetted some of these programs, looking at what data we would be putting out there, who owns it, what's the copyright status. Then we slowly started using it in schematic design and programming processes. We also began using it in image generation and color generation in our interior process."

Hoff says that MGA asks owners for certain words and prompts related to a specific space in their building, then feeds these terms into Midjourney's Discord interface – "warm, loving, caring" – and it produces multiple A.I.-generated images with a scene and a color palette complementary to those words. The program chose colors that fit the definition of the terms.

"We wouldn't just hand it over to the client without filtering it," says Hoff. "We use it behind the scenes. The owner wouldn't get those images. If we did give them those images, the best case is to give credit to the platform that created it."

ERROR DETECTION

Brian Burnett, senior consultant with PSMJ and former CEO of a 200-person engineering firm, says, "A.I.'s application in the AEC space can be around complex solutions. For example, it's incredibly useful around detection and error measurement to create a higher confidence in the plans being produced."

Burnett adds, "I've been gone from the company for six years, but we were using A.I.-type methodologies to perform error detection in our mapping work. During my career, we went from dozens of people looking for errors visually, to using the technology to detect errors based on selected search criteria. The concepts of A.I. have been around for a long time, but what's happened recently is that the market-readiness of A.I. reached a tipping point."

PROJECT MANAGEMENT

We have already touched on how A.I. impacts components of good project management – scheduling, cost estimating, collaboration, etc. A.I.-driven technology can also improve project management in areas such as resource allocation, scope definition and communication.

Architizer, an online product information and review service for design firms, named ClickUp the best A.I. tool for project management. "Adopted across many different industries, it has become a secret weapon, revolutionizing project management with features tailored for enhanced workflow efficiency. This cloud-based application seamlessly organizes and tracks projects."

Architizer says, "ClickUp harmonizes tasks, deadlines, and team assignments in a simple platform, ensuring project progression from design to construction, avoiding missed deadlines and maintaining momentum. ClickUp's document management facilitates effortless collaboration. Effective communication is key, and ClickUp delivers. Chat, video conferencing, and file-sharing tools synchronize efforts with team members and clients."

It concludes, "In ClickUp's virtual realm, real-time collaboration can be used to refine designs collectively, while ClickUp's reporting tools provide invaluable insights to identify improvement areas and fine-tune workflows. Analytics and visualizations offer panoramic project views."

This is one group's opinion and not an endorsement. Other A.I.-related PM software products include Newforma, Monograph, Archisnapper, Xledger, and Outplannr.

PRODUCTIVITY IMPROVEMENT

In 2017, McKinsey reported, "Globally, labor-productivity growth in construction has averaged only 1% a year over the past two decades, compared with growth of 2.8% for the total world economy and 3.6% in the case of manufacturing. In a sample of countries analyzed, over the past 10 years less than one-quarter of construction firms have matched the productivity growth achieved in their overall economies."

There's no indication that the situation has improved in recent years. In the United States, the Bureau of Labor Statistics said productivity in the construction industry grew from 2009 to 2013, but then steadily weakened through 2019. Productivity rose in 2020 and 2021, but declined in 2022."

Bartolf says, "One of the things that's sad about this industry is how bad productivity has been over the last 40 years. It's embarrassing. The only way to solve this problems is that we need to start optimizing parts of the building that we can prefabricate. Cut assembly time. Simplify the procurement process. That's what's going to improve productivity."

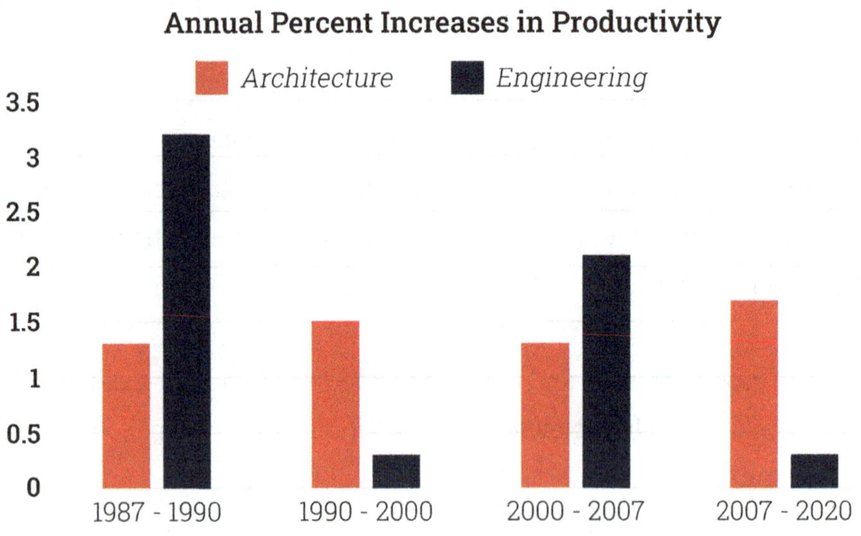

Source: U.S. Census Bureau

He adds, "I'd love to see A.I. be able to take a design plan and break it up to 1000 pieces that can be pre-manufactured. We've done that to a degree. At Gensler, we threw algorithmic design at the roof of the Nvidia headquarters. It's a swooping design of triangles, the code narrowed the number of unique panels in the design from thousands of options to just a few. If you throw machine learning – which is not A.I. necessarily – at a building, and break that building into a number of components, and deconstruct and reconstruct it like LEGO blocks, it could really go a long way to improving productivity."

VHB's Mulholland says the machine learning Deluge stormwater modeling program reduces what was formerly a five-step process to three. "Before, we'd take the lay of the land, see what it looks like, and do our design based on that. Next, we'd scan the land, start to do a design of the way the water flows, design it into the system, do all the calculations – the outflow and inflow – and produce the model. With Deluge, you do the base design to set the parameters, the sensors feed the data into the machine learning program, and then the iterative design takes over."

After decades of trailing other industries in productivity improvement, it might just be A.I. that finally gets the AEC industry over the hump.

CHAPTER 2 TAKEAWAYS

Here are the key takeaways from Chapter 2:

- The AEC industry is largely unprepared for A.I.'s impact on how we do business.
- The AEC industry will feel the disruption of A.I. more than most.
- Generative A.I. adoption is often a "bottom-up" process in AEC firms, but the C-suite is also driving much of the interest, investment, and strategy.
- In 2024, the reality of A.I. use in AEC falls short of "the art of the possible."
- Content generation – text and images – is primarily what AEC firms are using A.I. for as of early 2024.
- Most of the A.I. use seen in AEC today is focused on sales and marketing.
- A.I. use in accounting, HR and other back-office areas is growing slowly.
- Use of A.I. in project delivery lags that of other operational areas, but the potential is great.
- Vendors are developing new project delivery tools and enhancing existing ones, which will lead to vastly expanded use in a relatively short time.
- A.I. is already leading to efficiencies in the industry and could help improve or solve the productivity problem we've experienced for decades.

Chapter

03

WHAT'S NEXT FOR A.I. IN AEC?

What does 2024 and beyond hold for A.I. use in the AEC industry?

"I said early last year, in January of 2023, that it was going to be the year of A.I.," says Taylor Design's Steve Bennett. "[2024] is going to be about seeing how we can integrate it more into workflows. Taking what we've done so far and building on it. Creating a better foundation and putting it into play more. Maybe we can create our own system to use in our own network."

Forestay's Bruce Bartolf says entrenched external forces could impede AEC firms from broadly and rapidly adopting A.I. tools. "It's going to be interesting to see how it comes out," says Bartolf. "The industry is not architected well to utilize these tools. The AEC industry is an ecosystem with a huge number of controls around it. Code limitations, material limitations, building costs, site requirements, clients, weather – they all limit what we can do. If you set A.I. up against that, you may not be able to use it. The code guy might say, 'I've never seen this before, so you can't do it; it's not code.' There is this significant limiter on the edge of progress for the AEC industry."

Bartolf adds, "Another limiter is the .dwg format. What if I have a drawing that is parameterized? It's basically coded for a 3D printer. But the client who is accustomed to .dwg says 'I can't look at that. Give it to me in .dwg format.'"

PSMJ's Burnett stresses the importance of understanding A.I.'s possibilities within the context of a firm's technology and overall business strategies. "I'm a big fan of A.I.," says Burnett. "There's an element of the provocative about what can be achieved with it, but also a necessity to look at what's practical, and how A.I. fits into a firm's overall technology strategy. This spurs us to look forward to what's coming in the world of technology."

CONSOLIDATION OF A.I. TOOLS

One point of consensus among technology professionals is that there is almost certain to be substantial consolidation of the hundreds of A.I. tools currently available or in development. With this in mind, AEC firms should tread cautiously and tactically when investing time and money into A.I. technology.

"All anyone wants to talk about is the nuts and bolts of technology management," says Bartolf. "How is this going to affect us? But it's a bit early to make that determination. A.I. tools are very immature right now."

He adds, "What I tell people in AEC is that there's a cycle with software tools. Think about virtual reality and rendering. When you look at those technologies, there first were dedicated tools that weren't made in MicroStation or AutoCAD. There were all these small companies innovating with spin-off products. Then the big guys came along and bought up many of these providers, integrating them into the mainstream products. Now there's only one or two third-party rendering products out there. All the main players have it integrated."

Bartolf concludes, "I see A.I. in AEC progressing in similar ways. There's a bunch of little companies fighting among themselves with very targeted vertical tools designed to do specific processes. They'll either grow to become a fully functional suite, get bought out or go under. This will happen in the next three years."

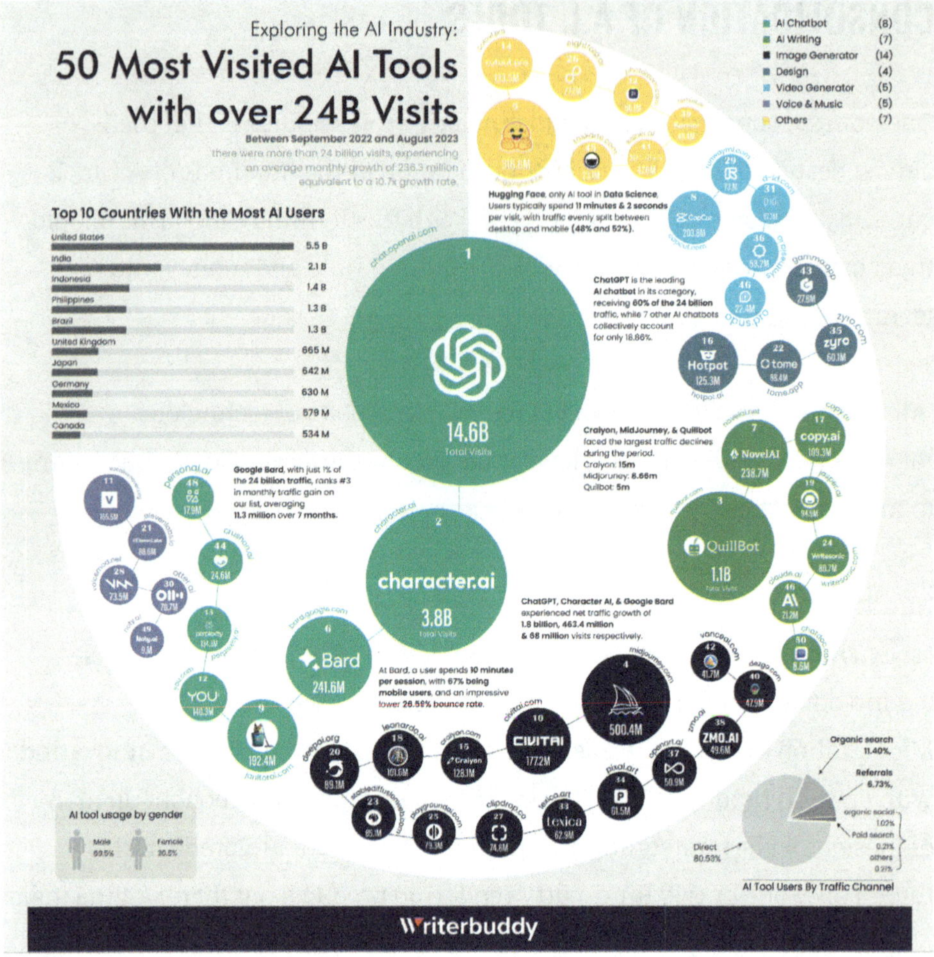

Lazaro agrees and gives the edge to the players with the biggest wallet. "There are literally thousands of tools out there. It's hard for even the people paying attention to understand what tools they should be using and need to be using. The ones that have the most money are going to stay at the top. The vast majority of tools are going to fail. Not because they're not good tools or they don't do fantastic or amazing things; they're just never going to get enough traction to make it to Main Street. Or more likely, they're going to be bought out by the bigger people."

This leads Lazaro to recommend the "name" brands, at least at this early stage of A.I.'s induction into mainstream business operations. "If you invest in companies like ChatGPT, Microsoft, Autodesk – all of those kinds of platforms – you're in pretty good shape because they're not going anywhere. A smaller tool may not live its lifecycle out."

The uncertainty that this likely consolidation engenders about A.I. tools suggests that AEC firms should move deliberately as they execute their tech strategy. Speed, in this case, could kill.

"In my role, I get 6-8 ads a day for different A.I. technologies, and 99% of those aren't going to be around [in the future]," says Mike Boisvert, Senior Consultant and Data Acquisition Manager for 1.000-person environmental consulting firm Verdantas. "If you're too quick to jump on a particular bandwagon, you're probably going to waste a lot of time and money. Having a strategic outlook is more important than moving fast because you may find that you've put yourself in a box. Focus on the long game, on scalability, and on whatever the technology needs [for it] to evolve and change for the better."

MULTIMEDIA A.I.

Although it served as a major disruptor and entryway into the use of A.I., ChatGPT was – and to a degree still is – limited to analyzing and outputting text-based data. Image-focused generative A.I. brought that sector much further along that it previously was. But the advent and expansion of A.I. products that can process and produce multimedia content within the same program could prove to be another tipping point for A.I.

"When Google launched Gemini, that really piqued my interest," says Unanet's Hayden. "What they introduced – and I'm sure others will follow suit – is multimodal processing. In a nutshell, with their platform, you can take different types of input, and process and produce similar varying types of output. We're moving from ChatGPT, where the input and output is text, to A.I. that can take in written, audio, visual and other forms of input. It can leverage all those mediums and then generate something that also leverages all of those mediums."

Upon introducing Gemini in December 2023, Google said that it "is the first model to outperform human experts on MMLU (Massive Multitask Language Understanding), one of the most popular methods to test the knowledge and problem-solving abilities of A.I. models. Gemini is built from the ground up for multimodality – reasoning seamlessly across text, images, video, audio, and code."

This type of multimedia capability expands the possibilities for business areas such as content marketing, proposals and presentations, predictive analytics, client deliverables, internal collaboration and many others.

Hayden adds, "Imagine an A.I. tool that can take in past project performance, proposals you've won, project imagery that you have in your system, drone video footage, client testimonial videos, press releases – you feed all those assets into it, and it aggregates those things and produces something really meaningful. Whether it's for another submittal, a presentation, fill in the blank. The possibilities are exciting."

Some critics say that Gemini is a step in that direction, but not yet ready to deliver A.I. tools to that promised land. MIT Technology Review's take on Gemini is reserved and a bit cynical, but not entirely negative. "It's a big step for Google, but not necessarily a giant leap for the field as a whole," write Melissa Heikkilä and Will Douglas Heaven in their review, "Google DeepMind's New Gemini Model Looks Amazing – But Could Signal Peak A.I. Hype."

"Google DeepMind claims that Gemini outmatches GPT-4 on 30 out of 32 standard measures of performance," they write. "And yet the margins between them are thin. What Google DeepMind has done is pull A.I.'s best current capabilities into one powerful package. To judge from demos, it does many things very well – but few things that we haven't seen before. For all the buzz about the next big thing, Gemini could be a sign that we've reached peak A.I. hype. At least for now."

In its earliest days, it appears that Gemini is a slightly better, slightly faster version of ChatGPT with some added multimedia capabilities. Future versions promise to further distance Gemini from its competitors, but time will tell if that promise is fulfilled.

"We're going to do things we haven't imagined," says Hayden. "We'll be able to produce co-creations that also output in those mediums. Creatives in our space will start to dream pretty big with the kinds of things we can leverage, both from the back office and business development. Also for project delivery."

PREDICTIVE ANALYTICS

Implementing A.I. for predictive analysis involves leveraging machine learning algorithms to make informed predictions about future trends or outcomes based on historical data.

"A.I. could provide a project score for an opportunity before you even make an investment to pursue it," says Hayden. "It can ask if it aligns with the company's current skill set or available resources. Have we done historically well on those projects, both from the client's and our own perspective? Or is this the kind of project likely to go way over budget? Maybe we have difficulty with this type of client. A.I. can allow a firm to take a more accessible look ahead, whether it's new pursuits or projects in flight."

Machine learning and other A.I.-related technologies have allowed predictive analytics to evolve quickly. An article in *Technology* titled "6 Top Predictive Analytics Tools for 2024," says, "The major shift in these tools is that it is getting easier to tune existing predictive modeling services or create new ones. Carlie Idoine, research director at Gartner, cited as an example her early (pre-Gartner) work in building predictive models to improve logistics. Her work then required deep knowledge of how the algorithms worked and also how to code them. Today, enterprises and vendors are exploring various types of augmentation to reduce the expertise and time required for many of the steps in this process."

Idoine said it doesn't require an expert to use predictive analytic tools anymore. "What used to require weeks of writing code can now be accomplished with a few mouse clicks and a lot of automation on the back end. The new capabilities of automated [machine learning] reduce the need to deeply understand how the variables affect each other, automatically choosing the best combination of algorithms for a given task."

The six products included in the *Technology* list are:

Altair A.I. Studio. The Altair RapidMiner platform includes a comprehensive set of predictive analytics tooling around its core data mining and text mining strengths, with Altair A.I. Studio supporting model development by data scientists.

H2O Driverless A.I. A relative newcomer to predictive analytics, H2O gained traction with a popular open source offering. The company's H2O Driverless A.I. simplifies A.I. development and predictive analytics for both experts and citizen data scientists through open source and custom recipes.

IBM Watson Studio. IBM became a leading predictive analytics tools vendor with the acquisition of Statistical Package for the Social Sciences (SPSS) in 2009. SPSS was founded in 1975 and grew into one of the top statistical and analytics packages over the years.

Microsoft Azure Machine Learning. Microsoft has long been a leader in various analytics capabilities through its Power BI analytics platform and Excel, which has become the analytics front end of choice for most business users. The company's Azure Machine Learning complements these core tools with capabilities for managing the complete predictive analytics lifecycle.

SAP Predictive Analytics. SAP Predictive Analytics is a good example of how enterprise application platforms can extend their core offerings to support predictive analytics workflows. The tool is a good choice for enterprises with an extensive SAP deployment, particularly those looking to create predictive analytics for logistics, supply chain and inventory management use cases.

SAS. SAS Institute is one of the oldest statistical analytics tool vendors. More recently, the company has modernized its core tool sets with various data science and machine learning workflows that take advantage of modern data stacks, augmented workflows and simplified deployment. The company has hundreds of tools for various domains.

This book's objective does not include product reviews, so we offer no recommendation on which of these – or the many other – predictive analytics tools are best for various tasks involving AEC firms. It's up to the individual firm to research and determine.

However, here are some steps that AEC firms can take to effectively use predictive analytics tools such as these:

Define the business problem. Identify the specific business problem you want to solve with predictive analytics. The more specific and well-defined your goal, the easier it will be to implement a predictive model to achieve it.

Gather relevant data. Collect as much relevant data as possible from all available sources. This data could include historical records, real-time data feeds, and publicly available information.

Clean and prepare the data. Ensure that the data is accurate, complete, and free of errors. This step is crucial since the quality and quantity of data play a critical role in the accuracy of the predictions.

Choose the right predictive analytics solution. There are many predictive analytics solutions available in the market. Choose the one that best suits your needs and budget. You can also build your own predictive models using machine learning algorithms.

Evaluate and validate the predictive mode. Test the predictive model to ensure that it is robust and accurate. Evaluate the model's performance and make necessary adjustments.

THE FUTURE OF A.I. FOR DESIGN

As much of an impact as early-stage generative A.I. is having in business segments like marketing content and image creation, the A.I. rubber will hit the AEC road when its use becomes second-nature in design and other core project tasks. As of early 2024, the industry wasn't there yet. When will it be?

"In AEC, like a lot of industries, Generative A.I. is really on the precipice of disrupting a lot of workflows and tools," says Lud Eng, Director of Information Technology for Stratus, a network of multi-discipline engineering, architecture, and consulting platform firms. "Early tools like generative design in Revit require a heavy upfront effort to get a good result. For this reason, generative design use remains limited because it takes a big upfront effort and level of expertise to get a project off the ground."

This is not likely to be the case for long, given the pace of change in technology for A.I. "It's easy to imagine a couple of generations of software down the line that it's going to become more user-friendly," adds Eng. "You won't need specialized training for design professionals to use those tools and they will take the drudge work out of the industry."

Unanet's Hayden says, "New design tools are coming out every day, and whether it's Autodesk or some other company, as products mature, you'll find more and more A.I.-driven features and functions as a way to help firms do business."

For example, in the June 2023 edition of the management newsletter The Friedman File, Verdantas' Boisvert said that A.I. has the potential to quickly determine the best approach to enhance job site performance. "Let's say we're designing earthworks for a shipping facility, and we have a set amount of material. A.I. may help minimize cuts and fills, identify where material should be stored for best functionality or to improve the flow of truck traffic. Clients aren't paying consultants for options or time; they're paying you to save them money in design. So A.I. might provide 15 options to decide from instead of 2 or 3."

VENDOR PROGRESS WITH A.I.

The rapid rate at which industry vendors are developing A.I.-related tools and enhancing existing products with elements of A.I. makes it nearly impossible to keep up. A thorough analysis of Vendor progress published one day is likely to be obsolete a month or two later. So here is a look at how some of the industry's top software vendors are approaching A.I. from a big-picture perspective as of early 2024.

AUTODESK

"Autodesk is putting a lot of resources into A.I.," Eng says. "They already have their first generation of A.I. tools out there, so it's not too hard to see where they're probably going to get to within a couple of generations. They have a great data pool for machine learning to train on, though some of the current tools are really on the border between generative A.I. or just a rules-based software system. But they're going in that direction. You don't need too many inputs for an A.I. program that pulls linear utilities along a roadway. Things like that will be pretty low-hanging fruit."

MGA's Hoff says, "Our current Autodesk reseller, IMAGINiT, has been creating some A.I. tools in-house, which they want to resell to us. We're a little hesitant. The cost is somewhat high; it can be quite a pill to swallow for our firm. Smaller firms can be at a disadvantage until the vendors figure out a price structure for us. To pay $50,000 for reseller access to something that saves us time, for a firm our size, may take years to pay off."

In a November 2023 blog post, "Evolving Your AEC Toolbox with Autodesk A.I.," Amy Bunszel, Autodesk's Executive Vice President, Architecture, Engineering and Construction Design, writes, "At Autodesk, we've invested in A.I. for the past ten years because we recognize its transformative power for the industries we serve. Across AEC disciplines and project types, Autodesk's advancements in A.I. are helping customers tackle complex challenges and harness new opportunities in their projects – not just by increasing productivity, but by giving them the tools to be even more ambitious and creative."

She says that Autodesk's focus is in three main areas:

- Analysis – immediate insights, earlier

- Augmentation – enhancing exploration and experimentation
- Automation – : reducing tedious tasks for more creativity

"Our solutions use a combination of A.I. technologies, mostly rule- and learning-based systems. No matter the type of A.I. technology, we believe it should always support the user, to help improve your workflows and optimize for the best outcomes. A.I. is simply a tool that's very good at completing specialized tasks."

ANALYSIS

Bunszel offers as examples:

- Forma's rapid analyses predicts results for noise and wind conditions and operational energy for early-stage design and planning, helping AEC teams design the healthy, sustainable communities they envisioned for their clients.
- Forma's rapid operational energy analysis gives immediate insights into a building's energy use, helping users design with energy efficiency in mind.
- With Autodesk InfoDrainage's Machine Learning Deluge Tool (mentioned elsewhere in this book), drainage designers can quickly generate responsive flood maps without having to rerun complex computational simulations. The tool analyzes and predicts water channeling and ponding, giving live feedback to help you complete a site analysis faster. This enables more informed decision-making in the drainage design process.

AUGMENTATION

Product examples include:

- Forma's initial set of generative design tools helps you answer the question 'what if...?' in a fraction of the time it normally takes when assessing a site.

- Generative Design in Revit gives designers a way to model the most optimal solutions by defining desired design outcomes, considered alongside competing goals and constraints.

AUTOMATION

Product examples include:

- AutoCAD, Markup Import and Markup Assist help drafters iterate faster.

- Markup Assist uses machine learning to identify markups on print or digital formats and help you incorporate changes easier and faster.

- AutoCAD Smart Blocks: Replacement suggests similar blocks from your block library for you to choose from.

- AutoCAD Macro Advisor generates macro insights based on your unique command usage.

In its review of November's Autodesk University 2023, *AEC Magazine* wrote, "From the get-go, Autodesk CEO, Andrew Anagnost, brought A.I. into focus, addressing the audience, 'You know that future that all of us thought was going to happen, sometime in the future? ...it is already happening. We've seen governments struggling to regulate it, schools are uncertain how to contain it, and Hollywood was going on strike over it. It is affecting our view of what's real. For better or worse, A.I. has arrived.'"

Anagnost continued, "And let's be really clear here. Done badly, A.I. absolutely leads to bad outcomes. Done well, it has the transformative power to understand and address the capacity issues that all of you face. The challenge of doing more with less – less labor, less resources, less money, and less friction – like the friction of moving your data back and forth between all the tools you rely on, that burns up the time and money and the people that get thrown at a project. A.I. can solve these problems and make your work more meaningful."

Anagnost called Forma, released in March 2023, "BIM reimagined." *AEC Magazine* writes, "Since its launch, Forma has seen a +500% increase in adoption. The product originates in the company's acquisition of Spacemaker in 2020. The team ported the conceptual design tool to the cloud where it's the first tranche of Autodesk design technology to be 100% cloud-based. While the original version was biased towards property developers, the program has since seen a lot of capabilities added to appeal more to architects."

The magazine adds that Bunszel said, "While today, Forma is aimed at early-stage planning Autodesk doesn't intend to stop there. Forma will expand to infrastructure and water. And development teams are building connections with Revit and Construction Cloud, with more to come."

Bunszel continued, "We're figuring out how we connect and build hybrid workflows between Forma and our existing products and what capabilities make sense to become Forma-native over time." Bunszel emphasized that Autodesk is building Forma to be open, interoperable, and A.I. powered to serve (the users), and that the most commonly requested integration for

Forma was with McNeel's Rhino, a commercial 3D computer graphics and CADD application software.

"You can bring your geometry from Rhino into Forma and can instantly analyze an impact of a design with factors like sun, wind and noise. And you can bring Forma terrain and surrounding building data into Rhino. This saves time and lets you work with your current tools," said Bunszel.

BENTLEY

"Bentley is working on some cool A.I. innovations," says Verdantas' Boisvert. "For example, they're offering a product that does good iterative design for site layouts."

In an article in *Technology* magazine, Julien Moutte, CTO of Bentley Systems, explains that the company "leverages data from IT, operational technology (OT), and engineering technology (ET) and compiles these three data sources in what it calls "intelligent infrastructure."

The article from December 2023, "Bentley Systems Turns A.I. Into an Infrastructure Assistant," quotes Moutte, who succeeded co-founder Keith Bentley in the CTO role. "It's more important than ever to make this infrastructure intelligence available to engineers so that they can do their best work. Because in our sector, infrastructure is under a lot of pressure right now due to project backlogs, talent shortages, and sustainability requirements."

The article continues:

"I think it's important for us to understand that A.I. and ML are already delivering considerable value to the infrastructure sectors. A lot of our accounts and users use A.I. today in their projects, whether they use it for something like computer vision to be able to assess the state of a bridge or a dam," says Moutte.

"Also, maybe to use A.I. to automate a lot of tedious tasks like looking at CAD drawings – thousands and thousands of CAD drawings and identifying the crosswalks – the shoulder widths and the lanes inside those designs to create that digital twin is something that is already happening today."

Moutte said Bentley Systems is seeking to address concerns of data usage and the reliability of A.I.—particularly in light of conversations where generative A.I. is at the forefront. Assurance is key to not only onboarding customers with A.I.-driven solutions, but also ensuring a thorough approach to infrastructure development, which means focusing on the transparent use of A.I. The company has committed to ensuring its A.I. solutions respect intellectual property and data privacy rights.

Moutte explains that, while A.I. has huge potential for utilizing data to create infrastructure designs, Bentley Systems emphasizes the need for A.I. to be a digital companion for infrastructure design, not a replacement for engineers.

"We believe this is an approach that is going to help our users understand how the data is being used, have control, but also make the best use of past data."

"Our tools already understand the rules of the physical world, the constraints of engineering, and we want to make sure that A.I. is working as an assistant to the designer and engineers and of our tools so that they can enforce those engineering rules and make sure we create safe designs."

At Bentley's Year in Infrastructure conference in Singapore in October, the company displayed what *ENR* called, "a robust digital twin of Singapore, with data from sources such as Google Earth and the Singapore Land Authority's own data collection and sensors, all collated and rendered in [3D creation tool] Unreal Engine. This view allows for deep dives into the nation's infrastructure and performance data, such as zooming in to a color-coded view of a given building's energy consumption, or seeing a real-time visualization of the country's mass transit system in action."

ENR said that a big topic of discussion at the conference was "the potential for greater use of subsurface geotechnical modeling in construction. Bentley acquired subsurface modeling firm Seequent in 2021 for $1.05 billion, and according to CEO Greg Bentley, the potential for bringing subsurface data into digital twins and project planning has barely been tapped by the industry."

Bentley is investing heavily in digital twin technology. It says that its " iTwin Ventures is a $100 million corporate venture capital fund that fosters innovation by co-investing in startups and emerging companies that are strategically relevant to Bentley Systems' objective of advancing infrastructure through going digital. The fund prioritizes seed to series B investments in transformational digital twin solutions supporting the design, simulation, construction, or physical infrastructure operations."

Tech media company FutureIoT reports that Bentley "estimates that the company's engineering users accumulate at least 100 million new unique digital components per month within their respective ProjectWise environments, teeing up potential infrastructure intelligence benefits across construction, operations, and maintenance. Among finalists to the Going Digital Awards, the use of Bentley Systems' iTwin Platform has risen to 64% in 2023 suggesting that digital twins are becoming mainstream."

DELTEK

Deltek claims 99% of the top 100 AEC firms as customers. Here are just a few ways Deltek says it is incorporating generative A.I. and traditional A.I. into its solutions:

Costpoint CRM. Deltek's market intelligence solution, GovWin IQ, is implementing generative A.I. into federal Agency Profiles and Contract Awards, in addition to SLED [state, local and education] Government Profiles with Smart Summaries. It is designed to transfer what was once a labor-intensive and time-consuming process into a seamless, real-time experience. Smart Summaries harness generative A.I. technology to provide clients with clear, concise executive briefings, eliminating the need for hours of manual research and data collection.

Smart Summaries are expanding into other data sets in GovWin, including Task Orders, Government Contacts and Opportunities to make understanding the background of key go-to-market elements consumable and shareable.

Costpoint PSA. GovWin IQ has extended its Smart Fit scores to all opportunity and contract data, revolutionizing how clients assess opportunities and quickly identify best-fit leads to pursue. With A.I.-driven Smart Fit scores, leaders can make informed pursuit decisions more efficiently, ultimately saving time and resources with suggestions based on historical contract awards. With visibility into competitors' or potential teaming partners' scores, contractors can also easily see who they might be going up against or who might be a strategic fit for their growth goals.

Costpoint User Experience. Deltek says, "When A.I. is combined with Deltek's Costpoint Business Intelligence (BI) solutions, government contractors can quickly collect, organize and analyze real-time data to inform their business decisions and better predict the success of a project. Deltek Costpoint BI uses A.I. to analyze data and generate a narrative around trends that can be used to support strategic planning or identify when a project is at risk or over budget – allowing you to course-correct. And with our A.I. Assistant in Deltek Costpoint BI, users can simply ask a question and receive a data-informed response quickly."

In December, Deltek said it planned to introduce its Costpoint Digital Assistant in 2024. The company says it "will allow users to be more productive and efficient. Powered by GenA.I., Deltek's digital assistant will permit users to ask questions regarding their contracts or projects and even carry out simple tasks like emailing a project manager with overdue tasks."

Software	Average Rating (5.0 is Best)		
	Small Firms	Mid-Sized Firms	Large Firms
BQE Core	2.7	2.0	•
Deltek Ajera	2.6	2.7	2.3
Deltek Vantagepoint	3.7	3.4	2.7
Deltek Vision	2.5	2.3	2.5
HSO Proserv AEC360	•	•	1.0
Unanet AE	2.8	1.7	2.3
BST Enterprise	•	•	3.7

CHAPTER 3 TAKEAWAYS

- AEC firms will build on the A.I. foundation they established with the release of ChatGPT and other A.I. tools in 2023.
- Inherent conditions in the AEC industry may limit how extensively, rapidly and well we adopt and use A.I.
- In assessing their A.I. strategy, AEC firms need to take into account not only what is happening now, but what is likely to happen with A.I.
- Consolidation of the hundreds of A.I.-related products in existence or in development is a near certainty.
- The consolidation will result in a few large players, most likely the ones who now possess the greatest financial resources (e.g., Microsoft, Google, Adobe, Autodesk).
- All-in-one multimedia A.I. products that build on text-reliant products such as ChatGPT are a significant advancement, though how significant is up for debate.
- Predictive analytics are likely to become a more widespread use for A.I. in areas such as go/no-go and profit projections because these tools have become easier to use.
- Plunging into A.I. for design work requires a substantial amount of upfront effort that many firms may not be willing to make, or that may not be financially or strategically worth making.

- The industry's major vendors are focusing many of their products on incorporating A.I., though there appears to be at least as much hype surrounding these developments as actual advancements.

Chapter 04

BEST PRACTICES FOR AEC FIRMS

Disruptive innovation -- aka "disruption" – is a common term for the impact that a new technology has on a business or industry. In the case of A.I., this disruption is likely to be monumental – possibly to an extent unlike anything in recent memory. As noted earlier, the impact of A.I. on the AEC industry is likely to be greater than most.

AEC leaders and technology strategists can prepare for the coming waves of change by understanding and implementing strategies, policies, procedures and other best practices, both for the new A.I.-related products that are already spreading throughout the industry and for tools and technologies in development or that exist only in theory.

In this chapter, we offer 13 Best Practices to follow to help ensure that you integrate A.I. into your firm in the most efficient and productive way possible.

REMEMBER THAT AEC IS A RELATIONSHIP BUSINESS

No matter what incredible devices and toys emerge from the A.I. revolution, it doesn't change the underlying fact that the AEC business is built on – and thrives on – the relationships among clients, service providers, team members, vendors and others involved in the overall process. When integrating A.I. into a firm's workflow, recognizing and respecting this dynamic must remain paramount.

"A.I. can bring productivity in newfound efficiencies to the table," says Hayden. "However, there are some best practices to keep in mind. For instance, when using A.I. for content creation, we should really think of this as co-creation because for the AEC industry, A.I. needs to be implemented in a relationship-tuned way. Relationships are in the DNA of this industry, and people can spot a fraud pretty quickly."

He adds, "Whether it's submittals, proposals, project status emails or something else, there's a lot of opportunity to communicate effectively with clients. Even project managers can turn to natural language responses; written responses that are nice and structured, and easy to read for the client. The key is to gain these efficiencies while maintaining a personal touch."

VHB's Mulholland says, "I have a lot of small firms call me and ask me to partner, so try to collaborate with firms that are like-minded, but may see some things differently than you. It's a relationship business, so know who you can trust. Participate in think tanks, and attend technology talks. Look for peer-to-peer groups. If you can find three to five firms and have peer-to-peer exchanges, it's amazing what you can accomplish in a two- or three-hour meeting."

Hayden adds, "Make your A.I. relational. Ensure that your brand, the voice of your culture and your firm comes through. It's a relationship business and the way you naturally communicate has to be preserved in any A.I. you might use."

ENSURE DATA QUALITY

You've heard it before, but it's true – garbage in, garbage out. Because much of the benefit we garner from A.I. comes from machine learning, the quality of the data that trains the technology will naturally dictate the quality of the output.

Unanet's Hayden cautions firms to assess and ensure the quality of their data before investing too heavily in A.I. "Ask if your data is A.I.-ready," he says. "A.I. is only as powerful and helpful as the data it leverages. If I'm a leader in an AEC firm today, I'm asking if our data is reliable, clean and trustworthy. If I leverage an A.I. tool, can I trust the output based on our data? A.I. adds opportunity, but with that comes a greater responsibility. That places even more importance on the quality of data."

Factors that Hinder a Company's Ability to Leverage Data

Select all that apply

		Tech Forward	Status Quo
Lack of sufficient tools to access, integrate, and analyze data	**41%**	35%	44%
Missing, outdated, or inaccurate data	**36%**	40%	34%
Lack of data availability or accessibility	**33%**	34%	33%
Little or no appreciation for using data	**27%**	24%	28%
Data isn't actionable	**19%**	20%	19%
Lack of resources	**1%**	0%	2%
None of these	**16%**	24%	13%
Other	**1%**	0%	1%

Source: Unanet AEC Inspire 2022-2023

While A.I. is reliant on the quality of data, it can also be a cleaning service. "AI can ensure that data is both accurate and enriched," says Hayden. "What I mean by that is, from a data stewardship perspective, A.I. is ensuring that we don't have bad data. For instance, that we don't have an 'Acme' and an 'Acme, Inc.' that are the same company. Data enrichment is important because A.I. provides the ability to enter a company in your CRM [Customer Relationship Management system] and have your system pull in information from LinkedIn and other public data systems to enrich a company record. It can provide important color and details to the data you're entering into the system."

Hayden adds, "You can use machine learning to take an email from Outlook, parse out the signature, and add it directly into the system's record for the individual. It can go through email in a client- or project-aware posture, subject to the data that's held in the CRM. Contact enrichment. Predictive modeling for forecasts. Financial automation. These are all examples of how A.I. is being used today."

Gauging the quality of data within the context of the firm's technology strategy and processes can help dictate which A.I. tool to use, Hayden says. "Ensure that your tech solutions fit neatly into your existing tech stack, so you not only provide consolidation within the stack, but it also provides an opportunity to be the hub of the organization's data and processes – like an organizational nervous system. A lot of firms, as they go through tech maturity, end up with several different point solutions. None of them integrate or talk together. They have duplicate data entries and manual processes. It slows things down and makes things less productive than they could be. If you have a solution that easily integrates with all technologies and synchs up the business in that way, it can result in major efficiencies for the business."

USE A.I. FOR EFFICIENCY

Early adopters of ChatGPT and other GenAI tools recognized that their greatest benefit lies in the ability to accelerate and automate a process. Rather than spend an hour musing over an employee recommendation letter, cover letter or some other time-consuming task, they learned quickly how to provide the A.I. tool with the appropriate prompts to deliver an output that would, at the very least, move the ball way downfield, if not all the way into the end zone.

Lazaro recalls, "I was asked to write a letter of recommendation for a former employee applying to an MBA program. Before ChatGPT, I would have spent an hour or more thinking about it, writing it, editing it. Instead, I told ChatGPT the former employee's best attributes, here's what I like about him, and then asked it to write a reference letter. And it came pretty close to what I would have written myself. I had to edit it some, but it took me 10 minutes total."

Hayden says, "At its foundation, A.I. adds new opportunity to the digital transformation or evolution that's already occurring for many AEC firms," says Hayden. "Whether it's moving off of paper invoices to doing something more digital and automated, or building out their tech stack and moving things to the cloud, A.I. adds another level of efficiency. It allows us to deliver better products within the space."

"Many organizations look at technology as a cost center," says Bartolf. "What they need to look at and understand is that with A.I., it's different. It's going to give you labor leverage. It's not a labor replacement, but it's one of the best tools developed in the last 50 years to help us automate the drudgery out of work. That means we're going to elevate people's work to only the higher-value things, while putting the lower-value tasks on the tools."

RECOGNIZE A.I.'S LIMITATIONS

A study out of Harvard Business School reports a major gap between levels of efficiency when an A.I. tool is used in the way it's meant to be used and when it's not.

MIT Review notes, "A new study on the impact of generative A.I. on highly skilled workers finds that when artificial intelligence is used within the boundary of its capabilities, it can improve a worker's performance by as much as 40% compared with workers who don't use it. But when A.I. is used outside that boundary to complete a task, worker performance drops by an average of 19 percentage points."

The paper's author recommends that rather than blindly adopting A.I. outputs, highly skilled workers need to continue to validate A.I. and exert "cognitive effort and experts' judgment when working with A.I."

BEST PRACTICE CASE STUDY: VERDANTAS

As ChatGPT and other A.I. tools were exploding on the scene during the first half of 2023, 1000-person environmental engineering firm Verdantas put a temporary hold on all A.I. use for deliverables. Speaking in the June 2023 edition of the AEC management newsletter *The Friedman File*, Mike Boisvert, Senior Consultant and Data Acquisition Manager for the 1000-employee environmental consulting firm, said, "We're establishing some guidelines within the company about the use of ChatGPT and other A.I. language systems. The current guideline is that no deliverables will be done with A.I. systems until we establish a vetting process. We want to create a task force to get a better understanding of what the system does before any published documents go out."

Boisvert added at the time, "We're taking the stance to not push this as a deliverable mechanism because we have no way of proofing where this information came from. Are there risks with submitting documents that might have proprietary or copyright information behind them? Where did the information truly come from? What facts are there? It's pulling all the information from various sources, so how do you properly source and document? It's a concern."

Yet, Boisvert was clear that he recognized ample opportunity for A.I. to transform the industry. "We want to use the technology," he said. "We want to engage Verdantas' employees to innovate and keep moving forward because we know that if we're not innovating and creating uses for A.I., we're going to be behind."

Six months later, Verdantas was in the process of rolling out its official policy, but use of A.I. was already advancing. As the firm continued refining its approach, Boisvert said in early 2024, "We want to build a foundational understanding of A.I. to develop comfort within the firm before we implement more advanced A.I. solutions. We envision using machine and deep learning solutions, and think that the best way to innovate with these technologies is through collaboration. Building a foundational comfort and understanding is important for creating an engaged community for creativity and ideation."

Boisvert adds, "We also need to be transparent. We don't want people to be afraid to talk about A.I. or use it in secret. The value of a shared learning process is figuring out the benefits, but also the weaknesses when it fails. This way we can build a community around it and integrate it into the culture."

Verdantas began the process by establishing an evaluation and integration

A.I. team. Each team member represents a different component of the firm. With Boisvert as the chair, the group includes representatives from IT, engineering, change management, project management and reporting, environmental data specialists, training and skills development, and other business units.

Boisvert says, "AI is a natural fit in a lot of ways, but everything we roll out – the system and solutions – needs to address a discipline's specific needs, and may require functional flexibility. It's important to roll out a system that's scalable and that addresses as many pain points or efficiency opportunities as possible, while evolving with innovation."

Ensuring broad representation is important, Boisvert says. "You could have a group of tech-savvy people who don't know what the A.I. use case might be for engineering, project management or IT. This is why we established the group; to have individuals who are subject matter experts in those areas to consider challenges A.I. solutions can assist with."

The group is diverse in terms of skill set and experience level as well, and ideas come from all areas. "We wanted innovative folks that cast the widest net," Boisvert says. "We have people sitting at the executive level, and we have people working in support and technology. Some of the best ideas came from people in the trenches. People who are doing the day-to-day project work are the ones coming up with solutions to assist with their pain points, so they can focus on the things that drive them. Additionally, having people from the executive side can aid in the evaluation of concepts and can engage their peers into the vetting process. This is why having a diverse team is the best way to get the most done."

Boisvert stresses that the unwavering support of leadership was key. "I have a lot of trust in the executive leadership team, and I believe they have trust in me," says Boisvert. "I came up through drafting and support, then worked my way up into leadership in a previous firm by creating teams to innovate and create beneficial change. When Verdantas brought me aboard, I told them that I needed the trust of the leadership team to be successful, and autonomy to put teams together in the same way. The A.I. evaluation group is an example of developing a community around initiatives that drive engagement and innovation. The executive team has been great about that request."

Verdantas released its A.I. policy around the time that 2023 turned to 2024. The firm created its policy with a third party consulting group, and passed its draft policy through legal and human resources experts before putting it into effect.

The key to the firm's policy is its decision to zero in on Bing's chatbot product "Copilot" as its base system. "We're using Copilot within our data securities policy," says Boisvert. "By keeping it protected, we address the concerns of privacy while sharing information. The system doesn't save information from chat conversations; when you close the window everything is erased. People can have a technical chatbot conversation, address the structure, tone and clarity of emails and deliverables, and help out with excel formulas and code."

Training to use A.I. is one area Verdantas emphasizes. "We wanted to roll out the policy along with a training program to get people comfortable with A.I.," says Boisvert. "Let's set up a foundation so that people can understand the value, look at use cases, and determine how to use A.I. as an assistant, not a replacement. We're adamant that A.I. is going to be used to empower and help our people, not replace individuals. That's high on our radar and it needs to be communicated."

Verdantas is using A.I. on tasks such as internal report reviews and document summaries. It is also assisting with project plan strategies, report formatting, image generation and marketing material. "We're not using it for technical documents yet," says Boisvert. "We don't use it to show schematics or anything technical-based. We're doing things like brochure imagery and animation. Adobe is pushing out new uses regularly for marketing and presentation assistance."

He adds, "The staff can use approved technology, but not to generate a report and send it out without detailed review and sourcing. It's like if you had an intern or early career staff person assist you with writing a report; you'd want to vet the information and verify that whatever they're compiling is correct, up to quality standards and sourced appropriately."

Part of Verdantas' A.I. approach is to brainstorm with staff. "We were getting a lot of requests, so we set up a system. People can request an A.I. solution they are interested in utilizing and explain what they want to use it for," says Boisvert. "The committee considers the submissions and tasks them out to the appropriate representative. The person assigned with researching it vets the idea, starts the process with any third-party vendors and reports back to the committee for next steps."

Boisvert advises AEC firm leaders to avoid straying too far outside their comfort zone with A.I. "If you use it as a partner, stay within the boundaries of your own expertise, and use it for the purpose it's meant for, it's going to perform faster and maybe better than you would without it. The risk is for people to use A.I. way outside of their wheelhouse, in ways they can't validate or totally trust. There's real risk with this, so we make sure to use A.I. only within our own comfort zone."

DEMOCRATIZE TECHNOLOGY INNOVATION

"Tech-Forward" is a term that has grown in popular usage to mean a person or organization that embraces and adopts new technology, and that uses technology to improve systems and processes. One way that a firm can ensure being tech-forward is to solicit input from and collaborate with a broad spectrum of employees. This is best accomplished with a structured program that encourages and facilitates the exchange of ideas. In this way, firms can identify and assess new tools with an eye toward their potential to help improve the firm's overall efficiency, operational performance and ultimate product.

VHB created an innovation program and opened it up to anyone in the company. The firm received 105 submissions and narrowed it down to 11 submissions to advance to the next stage. Mulholland says, "Once the submissions were identified, we scoped the projects and advanced them as investments."

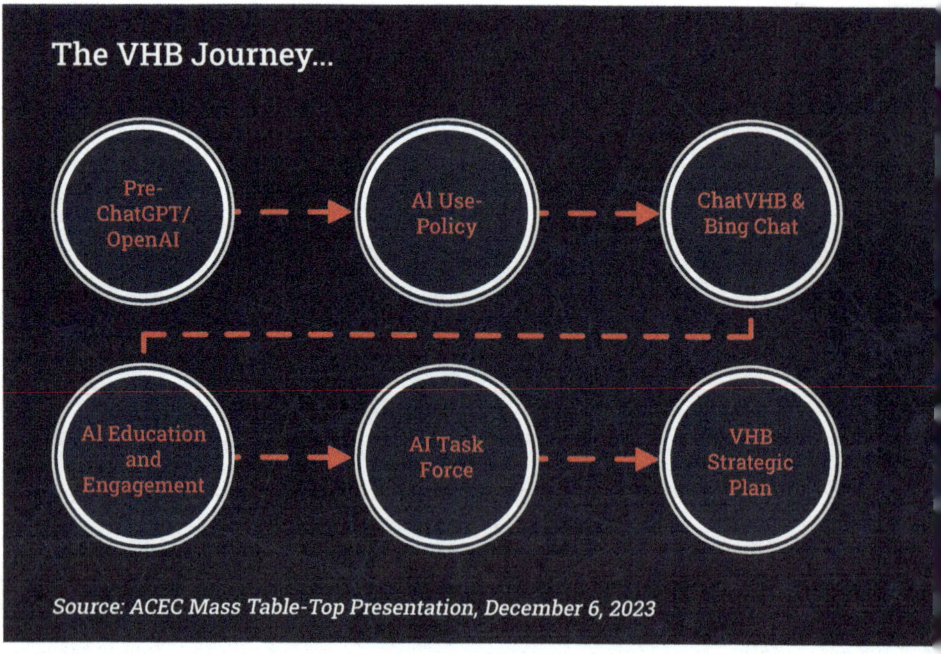

Source: ACEC Mass Table-Top Presentation, December 6, 2023

As an example, one of the selected submissions was a tool that would leverage machine learning to automate the evaluation of stormwater modeling. To inform the team of the advancements, VHB facilitated a livestream with the subject matter expert and technology lead. As a result of this livestream, the subject matter expert contacted Autodesk and found out that they already had a machine learning tool called Deluge. VHB wound up testing the Deluge tool for integration into the services that it provides to clients. As of early 2024, VHB continued to collaborate and partner with the vendor to further advance the project.

GIVE TECHNOLOGY A SEAT AT THE TABLE

Burnett says that AEC firms, in general, don't prioritize technology to the extent that's warranted by the importance of its role in the firm's operations, performance and success. "Technology is a huge opportunity for a firm to improve its performance in many areas," he says.

"[AEC firm leaders] need to understand that we're competing with technology firms all the time," says Mulholland. "If you don't embrace technology and make it an important part of your strategy, you're going to lose your competitive edge."

Bartolf says organizational learning is the key to capitalizing on new technology. "One of the things I've done in my career is focus on creating an organizational structure that supports innovation, but controls it in narrow ways," he says. "You need to have a client that's forgiving, a budget that's accommodating and a staff that's interested and supportive. This was Art Gensler's influence. We lost him not too long ago. He would identify a client and a team to pilot a technology, and specifically not hold that team to the same profitability number as other teams. They had to do a board-level report at the conclusion of a project about what they learned – what to do, what not to do, what needs more work. Art understood the value of investing billable time in learning how to do things better."

Median Values	Nonlabor IT Costs as % of Direct Labor	Ratio of Total Staff to IT Staff
Energy-Utilities	13.1	22.6
Staff Size 1 to 20	10.7	13.0
Staff Size 21 to 50	8.9	22.3
Staff Size 51 to 100	9.5	29.0
Staff Size 101 to 200	8.9	30.0
Staff Size 201 to 350	9.0	33.0
Staff Size 351 to 750	9.7	31.6
Staff Size over 750	9.7	43,6
Transportation	8.3	21.6
Water/Wastewater	7.5	31.1
Industrial	7.6	28.3
Commercial Users/ Developers	10.0	22.6
Education	10.6	25.3
Housing	8.3	24.3
Health care	9.5	29.9

PSMJ 2023 AEC Financial Performance Benchmark Survey Report

"I believe Gensler still operates that way, and has even added a formal R&D process to their work, publishing reviews of many projects annually," adds Bartolf. (You can find them at Gensler Research Institute – https://www.gensler.com/gensler-research-institute.) "If you want to invest in a technology, you have to apply to the R&D committee. If they accept the proposals, they'll give you money to support using the technology within a project or to do other research. They essentially subsidize the project. It could lose money. It's the theory of experiment and fail fast. They do narrow, rifle-type implementations affecting only part of the project. A lot of the time it's creating opportunities to experiment in areas that are of low value to the client, but essential to the completion of a project. Staircases for example."

Eng says, "As with most new tools, it's going to take some overhead time set aside for people to really get up to speed to see what makes sense for the organization to invest in, to position the firm to understand the tools as the they come out, and to understand how to get value out of them more quickly instead of playing catchup against firms that have already done it."

ENSURE THAT LEADERS LEAD

Burnett says, "You have to have leaders who believe it's a differentiator. You don't always have to be at the leading edge, but you can't be afraid to try things. You want to create an attitude in the team of hunting for new and better solutions. You also have to build your technology talent and expertise. I was blessed to be around people who could solve complex problems with different technologies – the COO who took over right before I left the firm was a computer science grad, not an engineer. It was a big move for us, but she was excellent at solving engineering problems with various technologies."

Burnett says that firms need leaders who "drive the process of integrating technology into solutions, and who have a vision of how technology can ultimately improve your firm's work. The only way to do this is to make it part of the strategy discussion like other parts of the business. You have to be proactive; waiting just puts you further behind. Ask lots of questions about what technology can bring to the table. Create relationships with vendors. We had strategic planning sessions twice a year and the topics were often technology driven. One of the most proactive steps is to question how technology can be used to improve project delivery, since this is the bread and butter of A/E firms. Actively explore how your processes and workflows can be improved, and what it takes to improve the skills of your staff."

SEPARATE TECHNOLOGY STRATEGY FROM IT

Firms that do best at using technology recognize the difference between the role that technology plays in the company's operations, workflow and product, and the IT function that keeps technology working smoothly and efficiently.

Bennett is charged with leading the strategic direction and use of technology at Taylor Design. "I help teams put technology into play in workflows," he says "When we're trying to spec something out, like the latest version of virtual reality for example, it's my job to make sure it works before I dump it in the lap of one of the designers. We also have a Director of IT who I interface with. This has been the trend for the last 10 years or more. To have someone to focus on BIM and related design tools and someone else to focus on the server side."

"Looking at the bigger picture, AEC firms need someone in the mix who understands the business implications of what these tools provide," says Eng. "It's going to change staffing. If you have work that will be automated by second- and third-generation tools, it would be smart to start cross training those people now so they can continue to add value as the tools evolve."

As CTO, Mulholland works closely with the firm's Chief Information Officer (CIO), but at VHB they are distinct roles. "Our CIO is one of my closest partners that helps support the infrastructure needed to support our technology initiatives." Mulholland says, "In a lot of firms you have either a CIO or a CTO. VHB strategically has both, and we partner to establish the data-informed, technology-focused vision, and ensure that it is implementable."

SIDE NOTES: "BRIDGING THE VALUE GAP"

In November 2023, A.I. and business strategist Elin Hauge was a guest columnist on the Deltek website. Her article, "Bridging the Gap Between A.I. and Business Value," details the ways in which the confluence of big data and advancing A.I. technology offer substantial opportunity for AEC firms.

"Fifteen years ago, big data was the hot topic. Then, about a decade ago, digital transformation came along...The natural next step is to apply mathematical algorithms to this data and intelligently analyze, assist, augment, and automate digital processes. And we call it artificial intelligence," writes Hauge.

She says three major changes have occurred: the enormous amounts of data being produced continue to grow and serve as training material for LLMs; processing capacity is greater than ever; and machine learning methods and architecture have improved significantly.

"Although A.I. may change the dynamics of the market in which you operate, you still need to keep your spotlight on the purpose of the company and avoid giving in to the temptation of diverting your attention to shiny new toys," she writes. "With that in mind, when you decide to leverage A.I. tools in your business, it may be wise to divide your strategy formation process into three parts:"

- Refine your core business with A.I. in mind.
- Grow the core with new products, services, business models, geographies, etc.

- Explore new business opportunities separate from core business metrics and objectives.

She concludes, "The bottom line is that strategy formation should take different paths for different purposes to balance operational excellence with radical disruption. The role of A.I. is ubiquitous, albeit some parts of the toolbox are more exploratory than others. Therefore, conscious considerations around the choice of A.I. tools may be of importance. A.I. is the natural next step in the digital transformation journey, and it is a toolbox with a wide range of capabilities. A.I. is, therefore, not a strategy but an enabler of your strategy."

https://www.deltek.com/sv-se/blog/ai-business-value

Mulholland says the pace of change for technology drives the need to have someone constantly considering the strategic perspective. "What is relevant today may not be relevant tomorrow," he says. "Six months from now, everything could be different. Firms need to be nimble. Flexible. Someone needs to focus on what's going on with technology outside the doors of the company."

FORM OR ENHANCE YOUR TECHNOLOGY COMMITTEE

The concept of a technology committee has caught on in the AEC industry, but many firms still lag in creating, supporting and ensuring the efficient function of a group that focuses on how the firm is – and should be – using technology.

Burnett wrote an article on the importance of having a technology committee for the June 2023 issue of PSMJ Newsletter. Here is an excerpt:

A/E Firms are Generally Behind

In my experience, architecture and engineering firms lag in the adoption and productive use of technology. In particular, the construction industry appears to be doing a better job of promoting opportunities to use technology as a tool. Maybe it's because our work is at the earliest stages of the creative process where technology applies to less of the work, compared with contractors who are focused specifically on optimizing the sequencing of how the design is best brought to life. Maybe we have so many options for technology that can help with so many things – from financial operations to project design to project management – that understanding how to apply technology is so much more complex. Or perhaps the speed of change with technological advances, especially as it relates to how data is collected and managed through the course of the design process, can be confounding compared to the built project that represents a static set of points, lines and curves.

Whatever the reason, we need to do better and go faster in finding ways that technology of all types can positively impact a firm's performance. How do we do this? In my strategic planning work, I stress the need for the topic of technology to have a bigger seat at the table. Technology needs to be a major part of the firm's strategy, just like business development for example, and the leaders driving the execution of this strategy must prioritize and track the progress of the firm's technology investments and implementation.

Step 1: Create a Technology Council

Many firms do this, but then let it wither away because of the daunting nature of the subject. Often this is because the wrong people are on the committee. Other times the group's leader and members don't have the skillsets to seek out, review and recommend new technology that fits the company's culture, work process, and financial needs. And, for many people, while technology surrounds us each second of our lives, we are uncomfortable with the responsibility of knowing exactly how to be a technological "change agent." But it is vital in today's rapidly moving business world to try, and try again, and try again. Unfortunately, there's no pill to make you a subject expert, so you must persevere through all the obstacles to make technology a bigger, more proactive piece of your firm's strategy, planning and operations.

Important Topics to Consider When Establishing a Company-wide Technology Group In an A/E Firm	
Technology Integration with Staff & Leadership Resources, Skillsets, and Strategy	**Technology-Based Systems and Processes**
• Integrating Technology with Strategic Planning • Understanding How Leadership Supports Technology • Recruiting for Technology Expertise • Outsourcing of Expertise for Strategy Creation • Outsourcing of IT Backbone Services • Creating Technology Expertise Redundancy • Training for Equipment and Production Software • Determining Competencies of IT Staff • Encouraging Innovation Ideas from Workforce	• Backbone System Hardware and Software • Data Management Requirements • Storage Requirements and Methods • Project Design Structure and Setups • Lifecycle Analysis for Equipment and Software • Means and Methods for Improved Efficiencies and Productivity • QA/QCc Processes and Technology Integration • Cybersecurity and Redundancies • Standardization and File Storage Framework
New Technologies and Advanced Strategies	**Technology Implementation**
• Advanced Visualization • Al and Business Development and Design Applications • Error Detection in Design • Drones, Robotics, 3-D Printing • Geographic Information Systems (GIS) • Digital Twins • BIM Strategies and Applications • Cloud Storage • Social Media and Technology Integration	• Technology Governance Approaches • Prioritization Methods for Technology Investments • Technology Planning and Budgeting • Organization of corporate Technology Talent • Technology Change Management Approaches • Computati on of Technology ROi • Licenses and insurance • Corporate Technology Policies

Step 2: Find the Best People to Lead and Participate

Despite the challenges that most firms face in using technology, some do it well. One that comes to mind is Pennoni, an ENR 500 consulting engineering firm with over 40 offices coast to coast. Their senior vice president and director of strategic growth, Joseph Viscuso, is a tireless advocate for spreading the message of how technology and innovation can improve the firm's client service, project delivery, and overall operations. The most successful technology committees not only have an inspired senior leader, but they also include a diverse and multi-generational selection of members from a cross-section of the firm's service and administrative areas. It's a committee that features people of different interests, backgrounds and experiences with technology that helps drive a comprehensive process to identify new products and technologies, review opportunities, determine potential use and return on investment, and so on.

Step 3: Ask the Right Questions

One way to ensure that a technology council doesn't ride off the rails is to consistently ask questions that demand an answer. For example, what is the current state of technology uses in the firm? Survey the staff. Identify the "Power Users" – i.e., the individuals who are the best at using a particular piece of software – for the major production software used by the company. What characteristics do these individuals possess that would be beneficial to transfer throughout the firm?

Other questions include: How can we perform better work with our major design production software through process improvements and training of the users? Where are we currently on the innovation adoption curve, and where do we want to be in the future? What is our technology inventory and replacement schedule? How are we perceived by clients and competitors for our ability to use technology effectively? What are the technology-related terms and words that need defining so that leaders understand? Does the company have a good inventory of the technological skillsets of the staff? How could we use A.I. efficiently throughout the company? Are we investing enough in technology training and budgeting?

Burnett also developed a matrix with ideas for how to best form a tech committee. (See table above.)

VHB is a strong proponent of technology-focused committees. Mulholland says, "We had 11 initiatives, which included model-based design. We established a team to advance the initiatives, but then ChatGPT came on the scene, and we had to accelerate our understanding of the large language learning model. With the advancement of A.I., we established an internal A.I. Task Force with three subcommittees. We needed to understand the impacts of A.I. on our core operational areas (e.g., facilities, human resources, finance, legal and marketing). We coordinated with the leaders of each of these groups who designated a point of contact to help advance the program and go back to their team to provide input and solicit feedback."

VHB also established a subcommittee to evaluate the impact of A.I. on client delivery, as well as a core technology subcommittee tasked with understanding partners and vendors and their respective position in the advancement of A.I.

ESTABLISH AN A.I. POLICY THAT WORKS FOR YOUR FIRM

Eng says, "You need a policy in the organization about how A.I. is used, and how that use is governed. A.I. systems can still confidently provide incorrect information, so it's particularly important to review it before it goes anywhere. You need everyone in the organization on board with the same understanding of where they can use the tools, and where they can't. Everyone must be clear on how to review and validate results before they impact a project or decision – well before a client would ever see them."

Just as Verdantas' Mike Boisvert slowed their use of A.I. until they'd had time to assess its impact and create a plan, Martin Gardner Architecture also got ahead of the A.I. curve.

"When ChatGPT came out, it opened our eyes to needing to be as early an adopter as we could be on some of the generative A.I. platforms," says MGA's Hoff, whose firm's A.I. policy is below. "We had a lot of junior staff, five or six years graduated, who really wanted us to be early adopters, and to get our arms around A.I. as quickly as possible. I, and others on our leadership team, said, 'Hold on.' We purposely slowed it down. We knew we needed to make some type of policy before some of the staff started using ChatGPT, Midjourney and some of the others out there."

Taylor Design is a 100% employee-owned firm, and they treat their staff like the owners they are. The firm's leadership gives people guidelines to follow, then trusts them to do the right thing. This is their approach to social media, which has resulted in exceptional, positive engagement, and the firm employs a similar policy for A.I.

"We released an A.I. policy to the company early [in 2023]," says Bennett. "We realized quickly that it was a hot topic that we needed to get in front of. We held a lot of open discussions and roundtable meetings. I went to each of our five offices, surveyed people. We settled more around principles of A.I. use, rather than hard and fast rules. It basically comes down to, 'Use A.I., but don't do dumb or harmful things with it.' It's principles like not sharing sensitive information with an A.I. program."

Bennett continues, "One of our core values is entrepreneurialism. We don't have the kind of rules that a large company might have. That's the only way to remain nimble enough that we can compete with larger firms that have more resources. A lot of our work is in healthcare, so there are obviously privacy concerns that you may not have in other types of work, but so far, we haven't blocked any sites to prevent people from using certain types of A.I. Even if we did, that isn't going to completely prevent people from using A.I. If people want to give information to get information from an A.I., they will find a way around methods preventing them."

The strictest guideline in Taylor Design's A.I. policy is to verify the accuracy of any A.I.-generated content in client-facing deliverables. "If we're using it in a proposal and use ChatGPT, we have to vet that it's correct. We want to be sure that the program hasn't produced any hallucinations (which occur when a large language model perceives patterns or objects that are nonexistent or imperceptible)," says Bennett.

The roots of VHB's A.I. Use Policy grew from the launch of ChatGPT in November of 2022. "The advancement of ChatGPT was not an obstacle for VHB, but rather an opportunity," says Mulholland. "VHB leaned into the large language learning model and the leadership at VHB became a big proponent of getting the message out. People need to be aware of the software and feel empowered to use the program to responsibly use the tool. In March of 2023, VHB's CIO developed a policy to create the guardrails for our team members that addressed issues like security, privacy, and confidential information."

Unanet's Hayden says, "Ensure that your firm has intelligence policies and procedures around how you'll use A.I. and what technologies your firm will support. It's especially important that AEC executives and leaders within the firm align on the policy as the firm starts to tread into the waters of A.I."

MARTIN GARDNER ARCHITECTURE – A.I. POLICY

Created by Amber Hoff, HR Consultant and President of Sound Mind Therapy, PLLC

GENERATIVE ARTIFICIAL INTELLIGENCE (GENAI)

Purpose: To ensure that the adoption and use of Generative Artificial Intelligence (GenAI) technologies within Martin Gardner Architecture are ethical, lawful, and in line with our organizational values and goals.

With the recent increase in use of GenAI, including programs like ChatGPT or Midjourney, it is important to stay up to date with the uses and benefits, as well as the potential negative consequences of utilizing GenAI in the workplace.

While we use GenAI on a daily basis, Martin Gardner Architecture has a duty to protect business and client information. Therefore, we are outlining guidelines regarding the use of GenAI in the workplace, specifically for work-related tasks.

Scope: This policy pertains to all employees of Martin Gardner Architecture, including independent contractors, who utilize GenAI for work-related tasks.

Principles: The use of GenAI at Martin Gardner Architecture should always aim to maintain transparency, accountability, fairness, privacy, and safety.

Guidelines for Images: Use of text-to-image generators and image preferences generation is allowed. However, images must be related to work tasks, and you must give credit to and cite the company or website that created the image to avoid copyright or plagiarism violations. Do not use photos of people (including employees at MGA or clients) to create images unless explicit written permission has been granted first.

"Definitely take your time; don't rush into things," says Hoff. "There are so many platforms out there, you have to find the ones that are right for you, especially when you have to pay for these platforms. Evaluate the process you have in place after you've done the research and see where it fits best in helping your processes. Once you figure that out, you want to write an A.I. policy around how you use A.I. in your firm. How you use it externally with clients, if at all. Be open-minded about all the different programs."

Lazaro recommends this basic framework to guide an A.I. policy:

1. Keep it simple.
2. Establish the dos and don'ts.
3. Identify what tools are acceptable to use.
4. Assess what software you are using today and what A.I. will be incorporated.
5. Ask what your customer are doing with A.I. Are they using it?
6. Ask if your customers have rules concerning A.I.
7. Ask what your training requirements are.

To create a more detailed policy, consider the following guidelines:

Establish Purpose and Scope. Clearly define the purpose of using A.I. within the organization. Specify the scope of A.I. applications and the departments or processes where A.I. will be employed.

Stress Ethics. Though "intelligent," A.I. is not human, and certainly not possessing empathy. Emphasize the importance of ethical A.I. use. Establish guidelines to ensure fairness, transparency, and accountability in A.I. decision-making. Stress the importance of identifying and mitigating bias in A.I. algorithms. Regularly audit and update models to address bias issues.

Address Legality and Compliance. Ensure that A.I. applications comply with relevant laws and regulations. Address data privacy and protection concerns, especially in industries with strict regulations (e.g., GDPR, HIPAA).

Address Data Quality. We established earlier the importance of having solid, clean and comprehensive data for the effective use of A.I. Outline data collection, storage, and usage policies for A.I. training. Emphasize the importance of data quality and security.

Encourage Transparency. Include guidance on when and what to say about the firm's A.I. use. Should any use of A.I. for any purpose be identified? Encourage transparency in A.I. models, particularly in explaining decisions to stakeholders. Consider providing explanations for A.I.-driven outcomes, especially in critical areas.

Focus on Security. Where the explosion of GenAI seemingly came out of nowhere, many users still don't have a firm grasp of what it is and what it does. This can result in substantial security risks, which is why many technology leaders are assembling guardrails around A.I. use. They're also regularly updating A.I. software and conducting security assessments.

Build in Training. Following on the previous point, it's the company's responsibility to train its people how to use the tools they're providing them (or at least allowing them to use). This is a critical factor, often overlooked.

Let the Benefits Flow. One common mistake in an A.I. policy – just as we saw in early attempts to regulate social media use – is the tendency to dictate rules without advocating the benefits. As previously mentioned, firms using A.I. to its fullest effect don't hamstring their people with endless rules. They focus more on what people can do with A.I., than on what they shouldn't do. They give solid counsel, sufficient information and strong motivation, then trust their people to do the right thing.

Communicate and Solicit Feedback. Establish a vehicle for feedback, collaboration, discussion and lessons learned. Continually monitor progress and keep the opportunity for dialogue active and open.

For a more in-depth look at A.I. use policies, including 43 actual examples from AEC firms, consult PSMJ's book, *2024 A/E Artificial Intelligence Policies*. You can find it here: *https://go.psmj.com/2024-ae-artificial-intelligence-policies.*

COMMUNICATE THE VALUE OF TECHNOLOGY TO CLIENTS

One of the dilemmas that firms face when it comes to investing in technology is the fear of receiving little to no return on their investment. It's one of the reasons that technology changes can sometimes be slow to take hold in the AEC industry; why spend money and time on a tool if it's not going to help the bottom line?

"A key is articulating the value of the tool to the client," says Burnett. "We were always trying to explain our technology solutions to our clients and other consultants, and how it would bring value to their work and solutions. Explaining the 'technology value' was very important to our success and application."

Here are a few ideas for how to communicate the value of technology to clients:

- **Tell it to them straight and trust them to understand.** If you explain the ultimate benefit of a new technology for the client's project success, while also explaining the investment required on your part, a good client – the type of client you want – will "get it."

- **Quantify the benefit.** The more detailed you are about how new technology aids the project, the more likely they'll understand. Educate them. Engage them. Answer their questions.

- **Don't talk them into anything.** Avoid the temptation to push a technology unless you're reasonably sure it will benefit the project and client.

DON'T GET CAUGHT FLATFOOTED...AGAIN

Maybe you were one of the rare firms that saw GenAI coming the way it did. (Do those exist?) Assuming you weren't, it's imperative to be prepared for the next wave. The best way to do this is by seeking out and understanding what's available now.

"This is coming whether we like it or not, so be prepared," says VHB's Mulholland. "Stay flexible. You're going to have to take calculated risks. Invest in technology that makes sense for your space."

Bartolf says, "In five years, we will have changed a lot of processes. You have to have your binoculars out there, wide and deep, learning about what's going on in the industry today. Grab a couple of products and implement them in small, targeted ways. Prove they work or understand how to change processes to make them work. Nothing works well unless you optimize the processes around the tools. Doing things the same way with new tools doesn't get you a great outcome."

BE DILIGENT WITH SECURITY

The cost of cybercrime continues to rise at an alarming pace. In 2020, the estimate was $6 trillion. Three years later, it is $8 trillion, a number exceeded in GDP terms by only China and the United States.

As happens when any new technology takes hold of the public interest, GenAI is a potential target for cybercriminals. Unlike a lot of emerging technologies, GenAI can actually play a role in combatting the scourge of cyberattacks.

"A.I. in cybersecurity, on both the offensive and defensive sides, is evolving rapidly right now," Stratus' Eng says. "It's an arms race, and it's probably going to be that way for some time."

Cybersecurity company CrowdStrike describes the pros and cons of GenAI in cybersecurity in its article, "Generative A.I. in Cybersecurity."

Efficiency. *With GenAI, cyber threat detection and response can become more efficient. As an A.I.-native system learns how to perform certain tasks, it can help security analysts surface information that they need to make decisions quickly. This accelerates analyst workflows, freeing them to focus on additional tasks, thereby scaling their team's output.*

In-depth Analysis and Summarization. *GenAI can enable teams to analyze data from different sources or modules, enabling teams to conduct traditionally time-intensive, tedious data analysis with speed and precision GenAI can also be used to create natural-language summaries of incidents and threat assessments, further accelerating and multiplying team output.*

Proactive Threat Detection. *Perhaps the most significant advantage of GenAI is the shift from reactive to proactive cybersecurity. By alerting teams to potential threats based on learned patterns, GenAI allows for preemptive actions before a breach occurs.*

Although the use of generative A.I. is compelling, it is important to consider the challenges that come with it. Like any technology, its use must be approached responsibly to mitigate risks and potential misuse.

High Computational Resources. *Training GenAI models requires substantial computational power and storage. For smaller organizations, this can be a limiting factor.*

Risk of A.I. Being Used by Attackers. *GenAI models and related tools become more accessible through open-source, inexpensive, and cloud-based means. Just as enterprises can leverage GenAI for cybersecurity, cybercriminals can use GenAI to develop sophisticated attacks that are adept at evading cybersecurity measures. Through a growing ecosystem of GPT-based tools, GenAI is lowering the barrier for new threat actors to conduct highly sophisticated attacks.*

Ethical considerations. *Many ethical questions of privacy and control over data, especially when it comes to the type of data used by A.I. models in training datasets, remain unanswered.*

The cybersecurity industry is struggling with a lack of qualified experts to meet the growing challenge of cybercrime. The small and midsized firms that comprise much of the AEC industry are particularly threatened because they may not have the resources to build sufficient defense against the malicious actors who are more than happy to attack them despite their size.

"The number of unfilled cybersecurity jobs is substantial, and the biggest gap is at the small and midsized company level," says Eng. "Larger enterprises, while they may have gaps, generally have a team of people working on cybersecurity. Small and midsized firms have maybe one person. With many cybersecurity jobs unfilled, a lot of companies have general IT people trying to fill in, and that's where there's a huge opportunity for A.I. tools to do a lot of heavy lifting."

Eng adds, "The companies building the tools have a lot of great talent. The big challenge seems to be having the right people to implement the tools at their companies."

Lazaro says he takes varying degrees of security measures depending on the situation. "Within ChatGPT in Settings & Beta, you can turn off 'chat history & training.' "This prevents anything you use from being saved or used by ChatGPT. The other option I am leaning more into is creating a MyGPT. This has several benefits, such as I can create one for each client, and it retains the data, but it is private only to me. It does take some effort to set up, though."

There will doubtless be more products and processes emerging to safeguard against privacy breaches and cyberattacks. Staying up on the progress in this field is critical.

"You have to prioritize cybersecurity, one way or the other," says Eng. "Budget to make that hire happen or find a good, comprehensive partner to work with who can provide those services. Not just a subscription service that checks email once a week. A full partner who will be able to work with the IT staff and provide monitoring, and respond to threats and attacks. There are a lot out there. Some are good, some are not. But this is really crucial because the threat actors are getting more and more sophisticated, and the environment isn't getting any better. You don't want to be trying to find the right cybersecurity partner after the attack has already happened."

Boisvert says, "I really like the idea of embracing the technology, but also understanding the risks behind that. Data security is a number one risk. That's why we're building a culture here that's comfortable with technology solutions, while understanding risk and working through the system approval process. We're encouraging people to think outside the box, and consider opportunities to utilize technology to help us work faster, more consistently, and with higher value for our clients, but also to understand the need for data security."

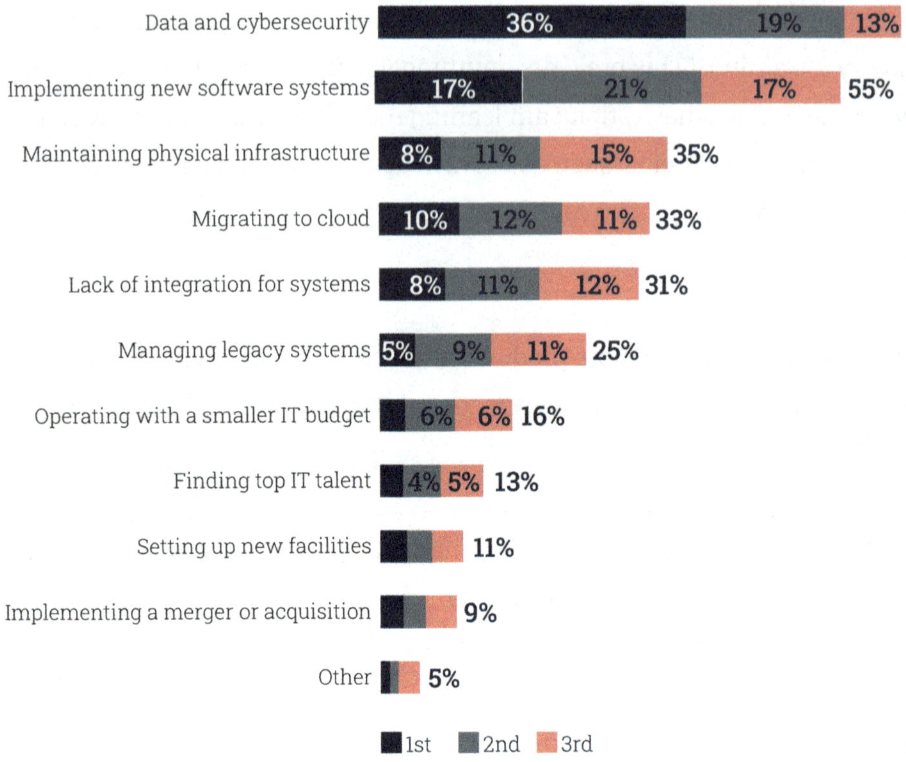

According to Deltek's 44th Annual Clarity: Architecture & Engineering Industry Study, more than 2/3 of A/E firms say that data and cybersecurity posed one of their top challenges for technology in 2023.

In October 2023, President Biden issued "Executive Order on the Safe, Secure, and Trustworthy Development and Use of Artificial Intelligence." It contains the following section:

Artificial Intelligence must be safe and secure. Meeting this goal requires robust, reliable, repeatable, and standardized evaluations of A.I. systems, as well as policies, institutions, and, as appropriate, other mechanisms to test, understand, and mitigate risks from these systems before they are put to use. It also requires addressing A.I. systems' most pressing security risks — including with respect to biotechnology, cybersecurity, critical infrastructure, and other national security dangers — while navigating A.I.'s opacity and complexity. Testing and evaluations, including post-deployment performance monitoring, will help ensure that A.I. systems function as intended, are resilient against misuse or dangerous modifications, are ethically developed and operated in a secure manner, and are compliant with applicable Federal laws and policies. Finally, my Administration will help develop effective labeling and content provenance mechanisms, so that Americans are able to determine when content is generated using A.I. and when it is not. These actions will provide a vital foundation for an approach that addresses A.I.'s risks without unduly reducing its benefits.

Issues That Keep You Up at Night

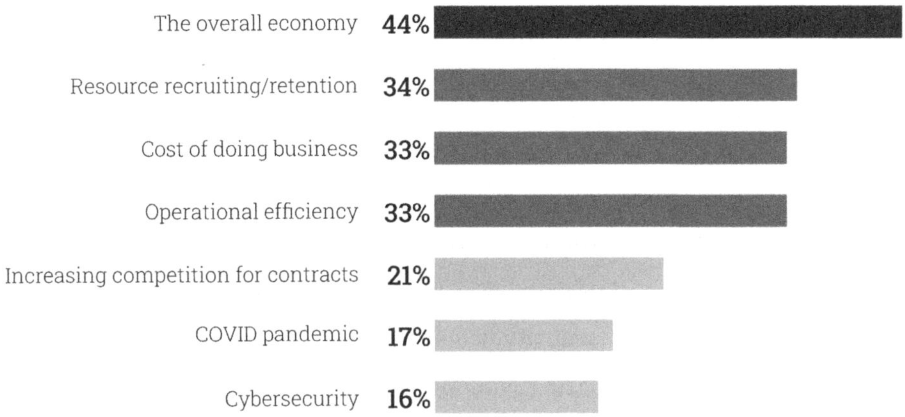

Unanet's AEC Inspire 2022-2023 report suggests that firm leaders aren't worrying enough about cybersecurity.

To protect A.I. systems from malicious attacks, unauthorized access, data breaches and other threats, consider the following:

Cover the Entire A.I. Lifecycle. Adopt a holistic and proactive approach that covers the entire A.I. lifecycle, from design to deployment to maintenance.

Implement end-to-end techniques and standards to protect, detect, correct and mitigate A.I. risks. This includes encryption, authentication, access control, auditing, testing, monitoring, debugging and incident response.

Design and institutionalize within the company the fundamental principle of data protection. Build safeguards into the A.I. systems to protect personal data, limit data collection and use, and ensure ethical practices. Aspects of this are required by certain regulations, including the EU General Data Protection Regulation, with much more likely to come.

Educate and train employees and partners on A.I. security best practices and policies. This can often be overlooked. Raise awareness, provide guidance and build a culture of security and responsibility among all stakeholders involved in the A.I. development and deployment process. Most firms already have best practices in place to promote these proactive protections.

Stay updated on security trends and regulations. This includes participating in industry forums, standards bodies and policy initiatives, and seeking feedback and advice from security professionals, publications, researchers and other sources. Hire experts if you need them.

CHAPTER 4 TAKEAWAYS

- With the pace at which A.I. and technology is changing, firms need best practices to stay competitive.
- Remember that AEC is a relationship business.
- Use A.I. for efficiency.
- Recognize A.I.'s limitations.
- Consult Verdantas' experience with A.I. as a case study for ideas on A.I. and technology.
- Democratize technology innovation.
- Give technology a seat at the table.
- Ensure that leaders lead.
- Separate technology strategy from IT.
- Form or enhance a tech committee.
- Establish a reasonable, meaningful and workable firmwide A.I. policy.
- Communicate the value of technology to clients.
- Don't get caught flatfooted by technology's rapid advancements.
- Be diligent about the security of your data in the age of A.I.

Chapter

05

MAJOR OBSTACLES & CONCERNS

A section in Chapter 1 addresses the challenges and considerations related to the functionality of A.I. In this chapter, we look more closely at a few of the major obstacles and concerns for AEC professionals and firms interested in using A.I. in their business.

WILL A.I. TAKE MY JOB?

The emergence of ChatGPT and other generative A.I. technologies revived this age-old question – is A.I. going to make my job or my career obsolete? It is not an idle concern.

The wonder – and the threat – of robots have hovered over humanity for centuries. Leonardo da Vinci is credited with the first recorded design of a humanoid robot, which occurred at about the same time Christopher Columbus was journeying to the Western Hemisphere. Robots have populated science fiction novels and films for nearly as long as the genre has existed, first appearing in a 1921 play by Karel Capek. The term derives from the Czech word for "forced labor." The first robots, as we know them now, were developed in the middle of the 20th Century.

In 2013, Carl Benedikt Frey and Michael A. Osborne published a report titled «The Future of Employment: How susceptible are jobs to computerization?". The authors examine how vulnerable 702 detailed occupations are to computerization. According to their estimates, about 47% of total US employment is at risk.

Using a similar method to this study, the website "Will Robots Take My Job?" offers an "automation risk probability" for each occupation. In 2021, the site began using BLS data, polling data and the automation risk probabilities to give each occupation a 'job score' between 0 and 10. The site creators determine the score by calculating job growth and availability, typical wages, and risk of automation. Higher scores are better.

For the occupation "Architects except Landscape and Naval," the calculated risk level is 0.0% and the job score is 6.7. By comparison, customer service representatives scored 82% and 2.2, respectively.

The website noted, "occupations in [the Architecture] category have a low probability of being automated, as they typically demand complex problem-solving, creativity, strong interpersonal skills, and a high degree of manual dexterity. These jobs often involve intricate hand movements and precise coordination, making it difficult for machines to replicate the required tasks." It cited important qualities of originality, social perceptiveness and negotiation.

In a poll of users, however, 34% of the 733 who answered said they believe the profession will be fully automated within the next two decades.

Civil engineers recorded the same risk level and polling numbers as architects, but its job score was higher – 7.8. The risk assessment was similar as well.

Mechanical engineers received 21% in the automated risk score, 34% among 2,020 survey respondents, and a 6.5 job score. The site says, "Jobs in this level have a limited risk of automation, as they demand a mix of technical and human-centric skills."

The consensus among experts about the AEC industry's susceptibility to replacement is that the following three statements most likely answer the question:

- In many disciplines within AEC – and perhaps most – A.I. is going to change the way you work at some point in time.
- For some functions in the AEC spectrum, the answer is actually "yes."
- For most others, the threat isn't from A.I. as much as it is from people who learn to use A.I. better and faster than you do.

VHB's Dave Mulholland was a featured speaker at a December 2023 ACEC Massachusetts Program, "Artificial Intelligence in the A/E Sector," which quickly sold out both the in-person and online maximum capacities. He closed with this: "People ask, 'is machine learning going to take my job?' The answer is, no, A.I. is not going to take your job. But those that embrace A.I. will have the competitive edge that will possibly replace you. At minimum, they'll have the competitive edge over you in the market."

Stratus' Lud Eng says that one way to judge which jobs are most susceptible is the length of time it takes for them to become proficient. "It cuts across disciplines," says Eng. "If you can take someone who is new to the industry and train them in a week or two in a certain workflow, that work is probably going to be automated in the future. Things that require a greater depth of experience will be automated later. For example, if you work in CADD doing linear utilities along a roadway, that may not be a career in a few years."

The key to overcoming the threat that A.I. presents is to treat A.I. as a tool, or an assistant, as Verdantas' Boisvert said in the Case Study from Chapter 4.

"Adding supplemental skills with A.I. is the power behind it," says Boisvert. "It's not replacing people; it's enhancing people. We're utilizing technologies to function better and with skill sets that supplement people's weak spots. I can use myself as an example. I'll spend a lot of time with email because I want to get it right, and make it clear and concise. A.I. is good technology to help me get it done quickly and accurately. It's improving a weak spot."

Lazaro says, "The biggest question that comes up is, 'will I lose my job?' It's a fear that a lot of people have. The simple answer is, no you're not going to lose your job to A.I. The reality is that you *are* going to lose your job to someone who is learning to use A.I. today. Mainly because they're going to be able to improve efficiency, make better decisions, and create ideas and innovation faster. And it's going to help enhance communication, whether it's in proposals, social media posts or even writing emails."

Will Robots Take My Job?

	Calculated	Poll of	Job
Occupation	Risk	Users	Score
Construction Managers	4%	22%	8.7
Civil Engineers	0%	34%	7.8
Architects	0%	34%	6.7
Mechanical Engineers	21%	34%	6.5
Surveyors	25%	42%	5.1
A&E Drafters	32%	51%	4.4
Construction Laborers	53%	55%	4.3

Risk of computerizaation in the next two decades for a selection of A/E/C occupations. Source: willrobotstakemyjob.com

SMALL FIRM DISADVANTAGE

Both conventional wisdom and common sense suggest that small firms face a handicap compared with their large-firm competitors in the race to implement and capitalize on A.I.

VHB's Mulholland says, "According to ACEC, nearly eighty percent of the companies in our industry have less than 100 staff members. But many software vendors require a minimum number of seats. These smaller firms will have a disadvantage compared to a 2,000-person organization."

Mulholland suggests that the competitive advantage that large firms hold in the race to incorporate new technologies may lead to even more consolidation in the AEC industry.

Taylor Design's Bennett says, "Small firms obviously lack the resources that bigger firms have. Gensler has staff to write their own application programs. Smaller firms have to look at products that are already published like Midjourney or ChatGPT. We have to rely on work that other companies have already done."

The disadvantage that small firms face is largely due to the following factors:

High Initial Investment Cost. Adam Tanton, in an article on B2B News Network titled "What Are the Greatest Threats to Small Business From A.I.?" writes, "The expenses associated with implementing A.I. technologies can include acquiring the necessary hardware, software, and infrastructure, as well as hiring skilled professionals with expertise in A.I. development and implementation. These financial requirements can pose a significant challenge for small businesses, especially those with limited budgets and resources."

Sometimes small firms need to consider whether an investment in A.I., even if possible, is a worthwhile strategy. As MGA's Hoff noted in the earlier discussion about vendor progress, small firms faced with the prospect of a major financial investment have to weigh whether the payoff in both time and profit justifies the expense.

"I've had conversations with a lot of partner firms who are less than 100 people," says Mulholland. "I'm finding that many firms want to do something with technology – they're hearing buzzwords and different terminology – but they don't have the means to do it. Bentley and Autodesk are the leading software vendors in digital delivery, and many firms are struggling with providing the necessary training programs to deliver projects effectively and efficiently using model-based design software. VHB has spent nearly 18 months developing a robust in-house training program, which a lot of the smaller firms can't offer."

Uncertain Need or Motivation. Stratus was formed through the private equity-backed alliance of five architecture and engineering firms – Prime Engineering, Austin Brockenbrough & Associates, Ascent Engineering Group, Morris and Associates, and Zyscovich. As of early 2024, it was using A.I. on a limited basis, but the circumstances and evolution of the tools had yet to drive widespread adoption.

"These were mostly smaller companies and the scope of most of their projects hasn't been conducive to the current set of A.I. design tools," says Eng, who spent over 25 years with Virginia-based engineering firm Draper Aden Associates before joining Stratus in 2023. "It's in its early stages at this company, so it's not an integral part of our workflow yet. We've had more adoption of A.I. on the corporate side – HR, marketing, even IT – using the more general tools that are already available."

“Some smaller firms, especially in our area, are so busy, they don’t have time to dabble with new technology,” says Hoff.

Less Robust & More Problematic Data. A.I. models need large and diverse data sets to train and perform well. Smaller companies may not have accumulated or have access to enough data – or the right kind of data – to train machine learning systems and build effective A.I. solutions. Smaller firms also may have fewer resources to ensure confidentiality, assess accuracy and organize data.

“Initially, small firms may be able to leverage A.I. tools as well as large firms,” says Bartolf. “However, large firms will find that the libraries of data that they have created will give them an advantage. Gensler has the floor plans for a majority of the Class A commercial real estate in the country. It’s a massive library of data. And what does A.I. like? Libraries of data. Large firms that have good libraries will be able to leverage that as an asset to develop higher-quality drawings and optimize those drawings.”

Lack of Technology Expertise. A.I. requires specialized skills and knowledge, such as data science, machine learning, programming and domain expertise. Smaller companies may not have the talent or the budget to hire or train A.I. experts. They may also struggle to stay apprised of the fast-changing and complex A.I. landscape.

Lack of Sufficient Infrastructure. A.I. requires high-performance computing, storage, and networking infrastructure to run and scale. Smaller companies may not have the capacity or the funds to invest in such infrastructure, or they may rely on third-party cloud providers that may pose security or privacy risks. In an October 2023 article, “10 Hurdles Companies are Facing When Implementing A.I.,” *Forbes* contributor Brian David Crane says, “Most businesses still struggle with outdated systems, tools and applications, making A.I. integration difficult."

Regulations. In response to a survey by the Small Business and Enterprise Council, 30% of respondents said that the use of A.I. needs government regulation. However, 84% of small business owners agree that government must strike a balance between regulation and fostering innovation "enabling the U.S. to lead in A.I. while safeguarding societal well-being," the SBE Council said in its October 2023 "Small Business AI Adoption Survey Report."

Some experts say small businesses are clamoring for more regulation in hopes of leveling the playing field, but others suggest that the opposite effect could occur. An article published on the Center for Strategic & International Studies website quotes Marc Andreessen (co-creator of the Mozilla Internet browser) as saying, "The development and proliferation of A.I. – far from a risk that we should fear – is a moral obligation." Andreessen calls the drive to regulate "irrational and hysterical fear," and criticizes big companies for "selling fear and pushing A.I. regulation to protect their economic interests. Big companies have the resources to meet extensive A.I. rules, while small companies and start-ups mostly do not," Andreessen said.

On a side note, the SBE Council is also demanding that small companies have more representation in the regulatory debate. "To date, the big company CEO's and academics have been consulted by Congress or received an invitation to the White House," the group said. "Small business owners need to be at the A.I. table. They are not only benefiting from A.I. tools, but they also dominate the tech industry and specifically this space. Congress, the White House and regulators need to hear from startup founders, entrepreneurs and small business owners in all deliberations about possible regulatory frameworks. Small firms are disproportionately impacted by regulation, and inappropriate or hasty regulation may actually mean the demise of important A.I. tools that could vastly improve access to health care and health outcomes, protect our nation, or help small businesses and their employees better compete and grow."

PRESIDENT BIDEN EXECUTIVE ORDER ON A.I. – JANUARY 2024

In October 2023, President Biden signed "Executive Order on the Safe, Secure, and Trustworthy Development and Use of Artificial Intelligence." Three months later, on January 29, 2024, the White House released a Fact Sheet on the progress made since the order went into effect, and reiterated some of its key components.

Agencies reported that they completed all of the 90-day actions tasked by the E.O. and advanced other vital directives that the Order tasked over a longer timeframe.

MANAGING RISKS TO SAFETY AND SECURITY

The EO directed a range of actions within 90 days to address some of A.I.'s biggest threats to safety and security. These included setting key disclosure requirements for developers of the most powerful systems, assessing A.I.'s risks for critical infrastructure, and hindering foreign actors' efforts to develop A.I. for harmful purposes. To mitigate these and other risks, agencies have:

- Used Defense Production Act authorities to compel developers of the most powerful A.I. systems to report vital information, especially A.I. safety test results, to the Department of Commerce. These companies now must share this information on the most powerful A.I. systems, and they must likewise report large computing clusters able to train these systems.

- Proposed a draft rule that would compel U.S. cloud companies that provide computing power for foreign A.I. training to report that they are doing so. The Department of Commerce's proposal would, if finalized as proposed, require cloud providers to alert the government when foreign clients train the most powerful models, which could be used for malign activity.

- Completed risk assessments covering A.I.'s use in every critical infrastructure sector. Nine agencies – including the Department of Defense, the Department of Transportation, the Department of Treasury, and Department of Health and Human Services—submitted their risk assessments to the Department of Homeland Security. These assessments, which will be the basis for continued federal action, ensure that the United States is ahead of the curve in integrating A.I. safely into vital aspects of society, such as the electric grid.

The original EO states, "The rapid speed at which A.I. capabilities are advancing compels the United States to lead in this moment for the sake of our security, economy, and society. In the end, A.I. reflects the principles of the people who build it, the people who use it, and the data upon which it is built. I firmly believe that the power of our ideals; the foundations of our society; and the creativity, diversity, and decency of our people are the reasons that America thrived in past eras of rapid change. They are the reasons we will succeed again in this moment. We are more than capable of harnessing A.I. for justice, security, and opportunity for all."

PRESIDENT BIDEN EXECUTIVE ORDER ON A.I. – JANUARY 2024 (CONTINUED)

INNOVATING A.I. FOR GOOD

The Fact Sheet continues: "To seize A.I.'s enormous promise and deepen the U.S. lead in A.I. innovation, President Biden's Executive Order directed increased investment in A.I. innovation and new efforts to attract and train workers with A.I. expertise." In the first 90 days after the E.O., agencies:

- Launched a pilot of the National A.I. Research Resource –catalyzing broad-based innovation, competition, and more equitable access to A.I. research. The pilot, managed by the U.S. National Science Foundation (NSF), is the first step toward a national infrastructure for delivering computing power, data, software, access to open and proprietary A.I. models, and other A.I. training resources to researchers and students.

- Launched an A.I. Talent Surge to accelerate hiring A.I. professionals across the federal government, including through a large-scale hiring action for data scientists. The A.I. and Tech Talent Task Force created has spearheaded this hiring action and is coordinating other key initiatives to facilitate hiring A.I. talent. The Office of Personnel Management has granted flexible hiring authorities for federal agencies to hire A.I. talent, including direct hire authorities and excepted service authorities. Government-wide tech talent programs, including the Presidential Innovation Fellows, U.S. Digital Corps, and U.S. Digital Service, have scaled up hiring for A.I. talent in 2024 across high-priority A.I. projects.

- Began the EducateAI initiative to help fund educators creating high-quality, inclusive A.I. educational opportunities at the K-12 through undergraduate levels. The initiative's launch helps fulfill the Executive Order's charge for NSF to prioritize A.I.-related workforce development—essential for advancing future A.I. innovation and ensuring that all Americans can benefit from the opportunities that A.I. creates.

- Announced the funding of new Regional Innovation Engines (NSF Engines), including with a focus on advancing A.I. The announcement supports the Executive Order's directive for NSF to fund and launch A.I.-focused NSF Engines within 150 days.

- Established an A.I. Task Force at the Department of Health and Human Services to develop policies to provide regulatory clarity and catalyze A.I. innovation in health care. The Task Force will, for example, develop methods of evaluating A.I.-enabled tools and frameworks for A.I.'s use to advance drug development, bolster public health, and improve health care delivery. Already, the Task Force coordinated work to publish guiding principles for addressing racial biases in healthcare algorithms.

For more information, visit ai.gov.

Not everyone believes that small firms are at a disadvantage. KP Reddy of Shadow Ventures says, "The 50-person firm is in a great spot to use A.I. to enable growth. Once you get to be a large firm, you don't want to shrink. Bigger ships are harder to move. I'm seeing some cool stuff coming out of 50-person firms, and it's deployed quickly. A.I. is leveling the playing field. With A.I., you don't need 50 people in a proposal department to compete with a firm that has 50 people in the proposal department."

It does make sense on one level that small and midsized firms that use A.I. to become more efficient and productive would improve performance exponentially more than a larger, slower moving firm might.

"Competing with the bigger guys for public work forced us to be early adopters," says Hoff. MGA recognized early how the efficient use of A.I. technology can give a smaller firm the appearance of being much larger than it actually is.

How can small and midsized firms ensure that they keep up with the big boys?

Tap into trade groups. Mulholland says that small and midsized firms should take advantage of training and other resources offered by their professional associations, like ACEC. Vendors are another good source of insight into where the industry is going. "Vendors are moving faster and investing more than what many AEC firms could," says Mulholland. "Firms can leverage the relationships they have with vendors through their licenses or software agreements to help them understand the mainstream tools available and prepare for the future."

Use federated machine learning (FedML). This allows multiple parties to collaboratively train a shared A.I. model without exchanging their data. Smaller companies can leverage the data of others without compromising their privacy or security. In its Fall 2023 issue, the *MIT Sloan Management Review* wrote, "FedML might turn out to be a game changer in addressing the digital divide between companies with and without big data and enabling a larger part of the economy to reap the benefits of A.I. It's a technology that doesn't just sound promising in theory – it has already been successfully implemented in industry."

Partner with vendors or consultants. Smaller companies can outsource some of their A.I. needs to external providers or experts that can offer customized or off-the-shelf A.I. solutions, tools, or services. This can save time, money, and effort. It can also help the firm access the latest A.I. technologies.

Invest in training. Instituting a program that educates A.I. users and shows them how to use the tools will result in a workforce more capable of capitalizing on A.I.'s benefits. It can also encourage innovation and enable better security.

Use low-code or no-code platforms. These are platforms that enable users to create A.I. applications or workflows without writing code or having technical skills. These platforms can simplify and automate the A.I. development process and lower the entry barrier for smaller companies. In a *Forbes* December 2023 article, "The 10 Best Examples Of Low-Code And No-Code A.I.," futurist and author Bernard Marr writes, "These terms are simply used to refer to tools that allow anyone to create A.I. applications without having to get their hands dirty writing technical code. A.I. can be useful to anyone in just about any job – from doctors and lawyers to marketers, teachers, and project managers. Many of these people will not have the technical skills needed to write code or the spare time to learn them."

The A.I. industry is increasingly developing tools that remove the need for coding skills and specialized knowledge to take advantage of the technology.

Content consultant Thomas Smith, CEO of Gado Images, says, "2023 was the year of the generalist A.I. model. LLMs like GPT-4 and Gemini are decent at performing a wide range of functions, from writing code to crafting Javascript. 2024, though, will be the year that generative A.I. gets specific.

More companies will create their own highly domain-specific A.I. models, or fine-tune existing frontier models for their own very specific use cases."

LIGHTNING-FAST PACE OF CHANGE

People don't agree on much these days, but everyone should be on board with the idea that technology is moving fast. If recent history is a guide, the pace will increase.

"The rate of change in the industry is accelerating," says Eng. "In the 1990s and early 2000s, the cycle of major technological changes in this industry seemed to be about five years. Then, through the late 2000s and into the early 2010s, it shortened to about three years. In the late 2010s, it seemed more like 18 months, and now it's maybe even heading towards 12 months. If your firm fell behind the curve ten or twenty years ago, you could still catch up in the same cycle. Maybe even in the first half of the cycle. Now the time between major change is getting shorter and shorter, and if you fall behind, it's increasingly difficult to catch up."

VHB' Mulholland says, "According to several studies, nearly twenty-five percent of all industries will be automated over the next five to ten years. As a practitioner, we should embrace the change, but not overreact. If you do nothing, you will be left behind. The industry is changing, whether we want it to or not. It is anticipated that our industry will change even faster than the general industry average."

Mulholland correctly notes that we've seen change in the industry before, but this time there is a difference. "I have a memo from VHB's Clerk on our Board of Directors issued in 1987 that addresses the transition from desktop drafting to CADD. Nearly 37 years later, you could literally replace 'desktop drafting to CADD' with 'CADD to model-based design,' and use that exact memo today. That being said, technology is moving faster in this industry than it ever has."

The speed of change is casting even more light on a longtime deterrent to progress in the AEC industry – chargeability. Too often, leaders want it all. They encourage employees to identify and pursue a strategic initiative of one type or another, but still expect them to hit their utilization number. On the other side, many employees and executives in AEC firms refuse to go an extra inch, never mind mile. So instead of a compromise solution in which the employee gives a few hours of their own time, but receives a minor reprieve from hitting their number, it's a standoff and nothing moves forward.

This is never a good situation, but in the Age of A.I., it could be deadly.

"There's a lot of pressure in this industry to keep billable time up, so people are working on deliverables because this is obviously how you make money today," says Eng. "At the same time, things are changing so rapidly and in such a seismic way that it is increasingly important for leadership to actively carve out time that's dedicated toward understanding these new tools and how they're going to integrate into the company."

The sit-and-wait philosophy may have been acceptable in earlier times, with different technologies. In many cases, waiting until the dust cleared and others made the mistakes was a viable, often successful strategy. No more.

"The pace of change in technology is a generational challenge within the A/E industry," says PSMJ's Burnett. "The reality is that firms get further and further behind due to how rapid things are changing. The only way to alter this trajectory is to proactively consider the opportunity that technology presents."

Burnett's firm benefited from technology innovation that moved them ahead of the competition by always challenging the status quo. In fact, early in his career, the firm created one of the first digital mapping platforms in the country. "Nothing was off the table when it came to exploring whether technology could solve a difficult or repetitive problem," he says.

"I've been through several cycles of different technologies," says Eng. "The leadership at my prior firm strongly supported investing in technology and recouping that investment. If you're in early, it gives you a way to recapture some of your investment before the technology becomes commoditized. Once it becomes mainstream, it's just the cost of doing business and doesn't provide a unique benefit to clients in the marketplace. Before, you could wait, make a smaller initial investment and let others absorb the R&D to figure out best practices. But at the current pace of change, if you wait too long it's already on the commodity part of the curve and there isn't a lot of room to earn the investment back."

Eng adds, "You'll need to spend some indirect time and some capital costs on the front side to set up and test new technologies. But if you get good at it early on, in the window where you're an expert on something that's providing a premium value for the firm or for clients, you have an opportunity to get higher fees of greater efficiency in workflow to recover the investment. If you can leverage a new tool to do in 10 hours something that traditionally takes the industry 50 hours to do, you have a nice advantage until the rest of the industry catches up. Or if you can do something that just wasn't feasible before a particular technology was available, and other firms aren't proficient yet, you have a great competitive advantage because you have a niche market all to yourself."

He continues, "We achieved both of those things a few times through strategic investment in innovative technologies. Of course, you have a limited window until the rest of the industry catches up and the workflow becomes commoditized, so you always need to be looking ahead to catch the next wave."

COPYRIGHT & INTELLECTUAL PROPERTY QUESTIONS

One of the least talked about, but important issues surrounding A.I. is how to deal with intellectual property. Concerns that legislation or regulations will likely need to address include:

Who owns the A.I.-generated content? There is no clear answer to this question, as different jurisdictions may have different rules and interpretations. In general, IP laws grant rights to human authors or inventors, but not to machines or algorithms. Therefore, A.I.-generated content may not be eligible for IP protection, or it may belong to the person or entity that provided the data, the input, the algorithm, or the infrastructure for the A.I. system.

Does the A.I.-generated content infringe on existing IP rights? Another concern is whether the A.I.-generated content may violate the IP rights of others, such as by copying, modifying, or deriving from their original works. This may depend on factors such as the similarity, originality, and creativity of the A.I.-generated content, as well as the source and quality of the data and inputs used by the A.I. system.

Is the data and input used by the A.I. system licensed and ethical? A third concern is whether the data and input used by the A.I. system are obtained and used in a lawful and ethical manner. For example, some A.I. systems may use data or input that are protected by IP rights, such as images, text, music, or code, without obtaining the proper permission or license from the owners. This may expose the A.I. system and its users to legal risks and liabilities.

Are we giving away too much of our intellectual property – or the wrong property – in the data we allow A.I. to access? A December article on the Educause website titled "7 Things You Should Know About A.I.," notes, "Generative A.I. tools are accepting larger amounts and more types of data, and it can be difficult to know whether the tools are retaining and using data for other purposes. This uncertainty risks exposing data that should not be shared outside of an organization, including personally identifying information and other private data. The ownership of both the data that the tools are trained on and the output is currently being litigated. Several lawsuits are pending against companies for using data without permission."

A Bipartisan Policy Center story, "Navigating the Era of A.I.: Implications for Small Businesses," by Michelle Kumar, says that A.I. technologies hold great potential for small businesses, but with concerns, including, "the security and privacy of the data used, both for small businesses and their consumers. To progress, A.I. systems require increasing amounts of data to improve utility and reliability. This creates privacy and security concerns over the amount of personal information available and fears around possible misuse – including bias against underrepresented groups. Policymakers need to address these concerns without stifling the growth of technology."

Some A.I. companies are trying to ease concern by including language in the agreement that absolves the user of liability in cases of libel or other illegalities related to the product. In November, Sam Altman of OpenAI said the company would defend anyone charged with copyright infringement if they used an OpenAI product. Microsoft, Adobe, Google and others have each made a similar pledge, though Bloomberg's Brad Stone, in a November 2023 article, "AI Legal Protections May Not Save You From Getting Sued," says it's not a guarantee.

WHAT AEC FIRMS SHOULD DO

In an article in the *Harvard Business Review* titled "Generative A.I. Has an Intellectual Property Problem," by Gil Appel, Juliana Neelbauer, and David A. Schweidel, the authors wrote, "Businesses should evaluate their transaction terms to write protections into contracts. As a starting point, they should demand terms of service from generative A.I. platforms that confirm proper licensure of the training data that feed their A.I. They should also demand broad indemnification for potential intellectual property infringement caused by a failure of the A.I. companies to properly license data input or self-reporting by the A.I. itself of its outputs to flag for potential infringement."

The story continues, "At minimum, businesses should add disclosures in their vendor and customer agreements (for custom services and products delivery), if either party is using generative A.I. to ensure that intellectual property rights are understood and protected on both sides of the table as well as how each party will support registration of authorship and ownership of those works. Vendor and customer contracts can include A.I.-related language added to confidentiality provisions in order to bar receiving parties from inputting confidential information of the information-disclosing parties into text prompts of A.I. tools.

"Some leading firms use generative A.I. check lists for contract modifications for their clients. Organizations that use generative A.I., or work with vendors that do, should keep their legal counsel abreast of the scope and nature of that use as the law will continue to evolve rapidly."

Using data that only comes from inside the firm has its own pitfalls. Bartolf says that implementing A.I. by tapping into a vast amount of in-house data could result in projects that tend to repeat aspects of a design. "The work a firm does may have a certain 'look' from using A.I. to mine their data libraries and propose solutions based on previous projects. The tools are going to create the look, because that's all they know. The stuff they make is going to look more like the stuff they made," he says.

CHAPTER 5 TAKEAWAYS

- Most positions in AEC are relatively safe from replacement by A.I., though occupations requiring little upfront knowledge or training will likely be fully automated in the near future.

- Professions and disciplines requiring greater levels of experience, dexterity and knowledge will avoid full automation, at least for a decade or two.

- AEC firms should look at A.I. as an assistant to people not a replacement for people.

- Small firms tend to be at a disadvantage compared with larger firms when it comes to investing in and using A.I.

- Key hurdles for small firms interested in A.I. include high upfront cost, a lack of need or motivation, limited amounts of quality data, lack of expertise, insufficient infrastructure, and regulatory issues.

- Some observers believe that small firms have an advantage because they can nimbly and exponentially improve efficiency and productivity.

- Small firms can counter any disadvantage by tapping into trade groups, using federated machine learning, partnering with consultants and vendors, and employing low-code and no-code platforms.

- President Biden's EO on A.I. attempts to put guardrails around issues related to A.I., though no worthwhile legislation is likely to hit as of early 2024.

- The rate of change in the A.I. universe is so fast that firms who fall behind may never catch up again.

- Firms need to compromise by giving A.I. champions ample opportunity to identify, explore and research new opportunities, while the champions likely need to be flexible with their time as well.

- The ownership of intellectual property, and the potential exposure of data that shouldn't be made public, are two concern related to IP when using A.I. technology.

Chapter 06

OTHER EVOLVING TECHNOLOGY ISSUES

In this chapter, we look at technology issues, trends and products impacting the AEC industry that are not typically associated with A.I.

TECH-FORWARD FIRMS

As noted earlier, tech-forward firms embrace and adopt new technology, using it to improve systems and processes. In its *AEC Inspire Report 2022-2023*, Unanet differentiates the firms it defines as "tech-forward" from "status quo" firms.

Effect of Pandemic on Recruiting/Maintaining Appropriate Staffing Levels by Technology Maturity

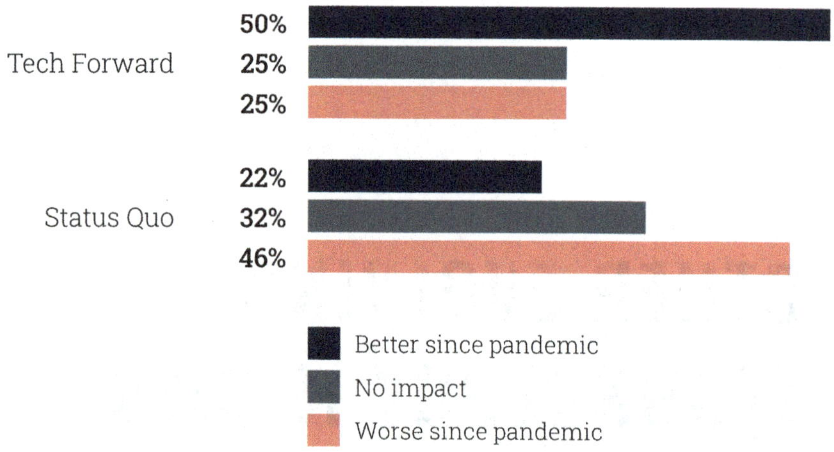

Source: Unanet's AEC Inspire 2022-2023

The report – which is subtitled *The Ascent of the Tech-Forward AEC Firm* – explains, "What exactly does being a tech-forward, data-informed AEC firm look like in practice? At the core, it's a firm that uses data and technology to support, complement, enhance, and empower people to work more efficiently, collaboratively and productively, make better decisions, faster, and ultimately, to build a sustainably profitable business that consistently delivers superior outcomes to its employees and clients. They also use technology to support their efforts to attract and retain talent, a major challenge in today's AEC business."

According to the report, less than one-third (32%) of the 395 AEC organizations represented in the survey qualified as "tech-forward."

"Firms that are technology forward are outperforming those that are technology static," says Hayden. "In our Unanet AEC Inspire Report, we saw the impact of this in a couple of areas. Profitability was one. Another compelling point was that since the pandemic, in terms of recruiting and maintaining appropriate staff levels, tech-forward firms report that metric as being better. Having a tech-forward posture impacts a lot of areas of the business."

According to *AEC Inspire 2022-2023*, "half of tech-forward AEC firms reported that recruiting and maintaining appropriate staffing levels actually have improved since the pandemic. Among tech-static firms, however, almost half (46%) indicated they have done worse in this area since the pandemic."

Hayden says, "In the war for talent, if a candidate has the opportunity to work for a firm that has its act together, that makes sure you have the technology to do your work with excellence and that makes the job easier, it can seal the deal."

The *AEC Inspire* survey also found that profitability levels for the prior 12 months and projected profitability for the next 12 months were both higher for tech-forward firms.

Actual Profit Rate Over the Past 12 Months

		By Industry			By Size			By Tech Maturity	
	Average*	Arch.	Eng.	Const.	10-99	100-499	500+	Tech Forward	Status Qou
	13%	8%	11%	17%	12%	17%	11%	15%	12%
	30%	26%	32%	32%	29%	24%	38%	35%	27%
	38%	49%	39%	38%	35%	40%	38%	36%	39%
	17%	17%	17%	12%	21%	19%	11%	13%	20%
	2%	0%	1%	1%	2%	0%	2%	1%	2%

30% 20% 10% 0%

PROFIT RATE

**excludes «Don't Know» responces*

Source: Unanet's AEC Inspire 2022-2023

Additionally, about one-quarter (24%) of tech-forward firms reported having no issues leveraging data; for other, tech-static firms, leveraging data is a major challenge. Tech-forward firms are defined in the report as being very data-driven, have 'fully or mostly integrated' systems and applications," rely on mobile capabilities and leverage cloud-enabled systems.

"You want to ensure that you're running modern technology such that you can leverage all the emerging capabilities in the market," says Hayden.

PSMJ's Burnett is also an advocate of the tech-forward approach. Here is an excerpt from an article published in the September 2023 issue of PSMJ discussing how firms can get there:

Technology is one of the most overlooked aspects in training programs for many AEC firms. Specifically, firms fail to consider how emerging technologies can improve efficiency, increase productivity and invigorate the project development process. Here are a few tips on how to ensure that your tech training program is functioning in a way that helps drive successful project management and delivery.

Regularly ask if there is new technology that can help you do something better. *Maybe it's once a quarter or before every project of a certain size, but firm leaders should take a step back now and then to investigate whether there have been any changes in available technology that can improve project performance. Especially for tasks that are repetitive or lack the need for creativity. Look at your next big project and consider how technology is going to be used, then see if someone has built a better mousetrap for some aspect of it since the last time you completed this type of project. If there is a new tool out there and time allows, get the best available people trained on it immediately. If they can't contribute their newfound knowledge to the current project, they'll be more than ready for the next one.*

Cross-train for work sharing. *A major limitation to good project management is the lack of versatility in technical professionals. Too often we encourage or allow specialization in our people, when both the person and the company would benefit far more if we had a team of generalists instead. In planning your firm's technology training, identify people whose existing abilities would translate best to related areas and broaden their skill set through technology training. It not only makes the individual employee more valuable to the firm and in their career, it gives leaders and project managers more flexibility in assigning resources.*

Train for consistency. *Inconsistency in the production process is a common area of concern for AEC firms. Many of the key people in a project have differing understandings of what is needed and how it should be delivered. Conducting training on the technology being used in a project can mitigate this problem by ensuring that everyone is on the same page in terms of the tools being used and how to use them. Equally important, the team will have a consensus on the way the data will be delivered and interpreted, and on what resources will be used and when. For example, in my former firm, we struggled to get our land surveyor subconsultants to deliver their data to us in a way that best suited our process. Had we planned around and discussed the technology being used, we could have had a more efficient project that was easier for everyone. Instead, we moved ahead too quickly and just assumed things would go the way we hoped. We were frequently disappointed that we had missed this opportunity.*

The firms that do the best with technology training invest sufficiently and plan strategically. They proactively identify specific technologies that will improve project and individual performance, then select the broadest, most appropriate group of employees to receive that training. They also recognize that technology is ever evolving, and they're not satisfied with keeping up with the competition. These are the firms that commit to being tech-forward leaders.

As part of a series on the Unanet blog, Hayden outlined some steps that firms can take to become tech-forward leaders.

He offers a step-by-step guide to becoming a tech-forward AEC firm and a few foundational elements to put in place:

Systems (ERP, CRM, etc.) that give you mastery over your data. Inside tech-forward firms, data and the insight it yields flow unimpeded and multi-directionally. This enables a firm to:

- Allocate resources more effectively
- Manage projects more efficiently
- Maintain a better handle on the KPIs that fuel project and firm profitability
- More closely monitor and manage the client and employee experience
- Strategically plan more effectively over various time horizons

Integrated systems. Your digital infrastructure should be fully integrated so your organizational "nervous system" runs on integrated apps, and your people are empowered in their work by automation, information and streamlined processes.

Cloud-centricity. Having key business applications and tools in the cloud offers scalability, efficiency and security advantages relative to on-premise solutions.

From-anywhere access to digital tools and data. With hybrid work environments likely here to stay in a business where employees often work in the field, it's essential that your people have ready remote and mobile access to the same digital tools, systems and data they have access to in the office.

Hayden then breaks it down by operational areas:

Accounting and finance:

- Implement stronger digital project accounting controls to avoid errors and delays in billing.
- Streamline invoicing and payments by automating AR and AP processes using integrated online payments, approval workflows, etc.
- Adopt digital accounting/finance solutions and ensure they are simple to use.
- Provide interactive analytic tools that make it easy for people to drill into and draw insight from data, and measure progress relative to KPIs (key performance indicators).

Project execution:

- Adopt digital tools that enable your firm to be more agile and proactive in managing projects, prioritizing a more flexible approach.
- Strive for collaborative solutions that enable people to work together to create positive outcomes.
- Automate processes wherever possible.
- Adopt analytics tools that provide live project insights so the right people understand exactly where things stand with a given project.

Capturing your market:

- Become more efficient and effective at developing proposals. An integrated CRM system with strong proposal-generation capabilities can help in this regard.

- Identify the client relationships that bring your firm the most value and, supported by digital relationship-management capabilities, work to strengthen them.

- Cultivate a collaborative knowledge "hive," where institutional knowledge is actively captured and readily available to people across the business.

- Develop digital capabilities to better predict your pipeline to inform forecasting, decisions about resources, etc.

With only about a third of AEC companies qualifying as tech-forward, and the results of surveys showing that these leaders in technology perform better overall, it seems that investing in technology is an opportunity not to pass up.

"As competitive as today's AEC business is, firms must be better than their peers at landing new business, attracting and retaining talent, and delivering superior project outcomes," Hayden wrote in his blog. "As roughly one in three AEC firms have discovered, an embrace of digital technology can provide that edge."

OPEN-SOURCE SOFTWARE (OSS)

"A growing trend is open-source software, which allows people to find their own technology solutions," says Burnett. "However it does take time to sort through the various solutions."

Open-source software is an evolving option in A.I., as discussed in the sections about predictive analytics and low- and no-code platforms. But its use is much broader.

As defined by application security company Synopsys, "Open-source software (OSS) is software that is distributed with its source code, making it available for use, modification, and distribution with its original rights. Source code is the part of software that most computer users don't ever see; it's the code computer programmers manipulate to control how a program or application behaves. Programmers who have access to source code can change a program by adding to it, changing it, or fixing parts of it that aren't working properly. OSS typically includes a license that allows programmers to modify the software to best fit their needs and control how the software can be distributed."

The benefits of OSS are that it is usually free or low-cost, it is relatively simple for people who are not technology experts to use, and it allows for customization. However, because it is either free or very inexpensive, the capability of the tools can be limited.

In the AEC industry, uses for OSS include:

Web development and design. For many smaller firms, this is often the first introduction to OSS. WordPress, Drupal and Joomla are open-source content management systems that enable easy creation and management of websites. WiX was the first Microsoft project to be released under an open-source license.

3D modeling, rendering, animation, and simulation. Blender is a popular open-source suite of 3D tools that can create realistic and artistic visuals.

Programming languages and frameworks. Python, Java and Ruby are widely used open-source languages that support multiple paradigms and platforms.

Databases and data technologies. MySQL, MongoDB and Apache Hadoop are open-source systems that can store, process, and analyze large amounts of data.

Operating systems. Linux is a dominant open-source operating system that powers many servers, cloud platforms, and devices.

Energy analysis and simulation. EnergyPlus, OpenStudio and Radiance can perform whole-building energy modeling, daylighting analysis, and optimization.

Benefits of OSS include:

Improving speed and innovation. Open-source software allows users to access the latest features and updates, and to collaborate with other developers and users. Users can also access the existing code and libraries to accelerate their development process.

Enhancing quality and security. Open-source software is subject to peer review and testing by a large and diverse community. Users can identify and fix bugs, vulnerabilities and errors more easily and quickly.

Attracting and retaining talent. The challenges and opportunities that go along with open-source software use can stimulate interest among existing and prospective employees. They can learn new skills, showcase their work and network via the tools. They can demonstrate their expertise, creativity and social responsibility by contributing to open-source projects.

VIRTUAL REALITY/AUGMENTED REALITY (VR/AR)

The American Institute of Architects (AIA) reported that 62% of architects have adopted VR or AR technologies in their design processes. Despite this impressive statistic, the prevailing viewpoint of VR/AR is that the technology is very cool and it has myriad uses and possibilities, but it has not yet reached its potential in the AEC industry. Vendors and experts talk a good game, and many projects are using the technology for marketing and project purposes. The question is whether the reality of these "realities" is on par with the hype.

"I'm waiting for that magic VR that everyone's happy with," says Taylor Design's Bennett. "Some people in the industry talk about the latest device that we need to be using – VR, 3D printing. Then I talk with designers fresh out of school and they say they'd rather just work with SketchUp or [Rhinoceros 3D] or whatever their favorite tool is."

Bennett adds, "No one has this overwhelming urge to get into VR. I wear glasses and most VR helmets don't fit, so I kind of lost any passion I had for it. Clients will occasionally ask if we do it and we tell them we do. But then once we put a price on it, they lose interest fast."

In 2018, a Gartner article titled "3 Reasons Why VR and AR are Slow to Take

Off," blamed poor 3D design interface, the lack of convenience and control in VR, and competition from smartphones.

More than five years later, these questions remain valid. Is a virtual reality walkthrough using an expensive headset that much more compelling than a 3D walkthrough on a screen (that more than one person at a time can view)?

A report by Sara Rankohi, Mahsa Rezvani, Lloyd Waugh and Zhen Lei, *Augmented Reality Application Areas for the Architecture, Engineering, and Construction Industry*, states, "Augmented reality (AR) is among the technologies that have the potential to advance the Architecture, Engineering, and Construction (AEC) industry. Yet, studies show that there remain challenges in applying AR in AEC."

The study report, which is based on the Pierre Boulanger book, *Applications of Augmented Reality - Current State of the Art*, says, "Various applications have been identified as visualization and simulation; in-situ experience; real-time information retrieval; maintenance, inspection, and repair; project documentation; heavy equipment operation; educational training; health and safety; site navigation; and automated measurements. To further explore these application areas, a case study was conducted using the AR solution of Trimble XR10 with HoloLens 2 in a precast construction context. The results show that existing AR technologies and systems for simulation/visualization and construction quality control are still immature. The study highlighted the current use cases, the potential for technology improvements, and the obstacles that hinder the widespread AR implementation in the AEC industry. Considering these factors, further directions and future research paths for innovators are proposed."

Though the study focused on AR, it is also applicable to VR. The authors identified the following challenges of using AR in AEC projects:

High initial cost and low return on investment. Like many technology investments, ROI can be hard to come by, the authors concluded. "Augmented reality (AR) implementation in construction requires substantial investment in hardware, software, training, and maintenance. Construction companies need to carefully evaluate the cost–benefit ratio and calculate the potential return on investment. Demonstrating the tangible benefits and long-term cost savings of AR technology can be challenging but is essential for wider adoption. Since AEC projects are typically large and complex, a prevailing notion indicates that applying AR tools and technologies would require a significant number of expensive devices, such as HMDs and other related equipment. Subsequently, a major initial investment would be necessary to integrate AR into construction and infrastructure projects, which can expand the overall project cost."

Nascent technologies. The authors write, "Augmented reality (AR) technologies are in nascency in the AEC industry. It requires powerful hardware and infrastructure to deliver real-time, high-quality AR experiences. Construction sites are often remote and lack stable internet connectivity, making it challenging to provide the necessary infrastructure for AR implementation. Additionally, AEC projects demand a higher degree of precision, consistency, and efficacy. However, current AR devices struggle to control the extremely sophisticated 3D information models regularly used in AEC projects. Originally, AR devices were designed for the entertainment industry, thus, they might lack the on-site capabilities necessary for the AEC industry. In addition, due to the complexity of this technology and a lack of technical skills and awareness, AR adoption in the AEC industry is slow."

Insufficient adoption demand. "Studies show that the demand for AR in the AEC industry is currently low. Nonetheless, with the ongoing evolution of technology and the formulation of customized answers to address the distinct requirements of the AEC sector, there's an anticipation of heightened acceptance. To foster its adoption, initiatives aimed at raising awareness should be launched, highlighting the benefits of AR technology within AEC. These advantages encompass enhancing project comprehension, optimizing cost-effectiveness, and enabling efficient training. Furthermore, educational establishments can contribute significantly by enticing fresh expertise in the field through the introduction of programs that offer specialized training in AR," the study report states.

Lack of experts and insufficient training. The authors write, "Introducing AR technology requires training and upskilling the construction workforce. There may be resistance to adopting new technologies due to a lack of awareness or skepticism. Overcoming the learning curve and effectively training workers to use AR tools can be a challenge. A limited number of individuals are pursuing careers in AR due to its immature stage of development, complexities in adoption, and absence of standardized implementations. This lack of established norms complicates the evaluation of the knowledge and skills possessed by those working with AR technologies. Moreover, the scarcity of experts in this field poses difficulties in gauging their grasp of the technology and effectively involving them in financially significant AEC projects. To tackle this challenge, universities should introduce advanced education programs centered on AR, thereby bolstering research and development endeavors, and enhancing the technology itself. Concurrently, AR companies could support research and development initiatives within universities, thereby facilitating this progress."

Poor user experiences. "Extended utilization of AR devices like HMDs can result in motion sickness, queasiness, perspiration, headaches, and even vomiting among users. This poses a significant impediment to the widespread adoption of AR technology. Given the intricate and time-intensive nature of AEC projects, prolonged engagement with these devices can prove distressing for most users. To address these concerns, endeavors should be directed toward enhancing the design and development of AR devices, aiming to minimize discomfort and alleviate the adverse effects experienced by users. Additionally, incorporating frequent intervals during extended usage and restricting the duration spent in the virtual environment can aid in diminishing the occurrence of motion sickness and other related discomforts," the report states.

Integration, adaptability, and data security. Integration with legacy systems is a common problem in the current state of the industry. The authors write, "Integrating AR seamlessly into existing construction workflows is crucial for adoption. AR systems need to interface with existing project management software, BIM data, and other AEC tools. Ensuring compatibility and smooth integration can be a technical and logistical challenge. The AEC industry involves multiple stakeholders, each using different software and systems. Achieving standardization and interoperability among various AR platforms, software, and hardware devices is essential to enable seamless collaboration and data exchange. AR systems in construction involve capturing, processing, and sharing sensitive project data. Ensuring data security, protecting intellectual property, and addressing privacy concerns are crucial. AEC companies must have robust data security measures in place to protect against unauthorized access or data breaches."

The authors conclude, "Addressing the challenges...requires a collaborative effort between AEC companies, technology providers, regulators, and industry associations. As AR technology continues to evolve, these challenges are likely to be mitigated over time, leading to more widespread adoption and integration of AR in the AEC industry."

It does seem feasible that as the technology improves, particularly as A.I. becomes a greater part of the picture, the use of VR/AR will grow.

On the website of AEO Design & Construction, Kamaljeet S. Marwah writes, "With an annual growth rate of 113.2%, AR, VR, and [mixed reality] markets are expected to hike and reach $250 billion by the year 2028. One of the significant transformations in the landscape of the AEC industry was brought about with the introduction of BIM. As technology advances with the latest inventions, AR and VR paved their paths in the construction sector as well."

In the December 2023 article, "The Future of AEC with VR and AR," Marwah lists the following benefits that he expects will grow VR/AR use in the AEC industry, along with the reasons behind these beliefs:

Project Management. With clash detection and instant integration of revisions, project management becomes a breeze, and the scope for delayed communication gets minimized amongst different teams.

Minimized Time and Maximized Productivity. AR can be adapted to support site work as the virtual model allows the detection of collisions which cuts down the need for rework. This proves especially useful with Revit MEP modeling and structural details.

Lesser Risks in Construction. The safety asset is improvised by visualization with AR and VR which cuts down the on-site construction risks.

Enhanced Collaboration. Most of the design issues arise due to a lack of communication. AR- and VR-enabled Revit MEP modeling helps in better collaboration with one single model.

Enhanced Client Communication. The easiest way to make your client understand the ideas is through a 3D model. While VR offers them a virtual walkthrough of the design, AR depicts its relationship with the on-site surroundings.

Collision Control and Bridge Maintenance. An in-depth study can also be conducted through Revit's structural modeling to work on collision control. It can also be adopted for bridge maintenance through point cloud data.

An Immersive Experience. AR and VR add value to the construction industry as these offer an immersive experience in the virtual environment before construction takes place in the physical world.

Some of these uses are occurring now, though not to the extent that advocates may have hoped. It remains to be seen whether there will be the type of breakthrough in the industry that Marwah predicts. For now, however, VR and AR appear to be niche products for much of the industry.

3D PRINTING

In 2015, the *Harvard Business Review* announced "The 3D Printing Revolution: It's Happening and It Will Transform Your Operation & Strategy." Almost a decade later, has it?

Here's how the article by Richard D'Aveni begins, "Industrial 3-D printing is at a tipping point, about to go mainstream in a big way. Most executives and many engineers don't realize it, but this technology has moved well beyond prototyping, rapid tooling, trinkets, and toys. "Additive manufacturing" is creating durable and safe products for sale to real customers in moderate to large quantities.

"The beginnings of the revolution show up in a 2014 PwC survey of more than 100 manufacturing companies. At the time of the survey, 11% had already switched to volume production of 3D-printed parts or products. According to Gartner analysts, a technology is 'mainstream' when it reaches an adoption level of 20%."

Only a year later, PwC said that 3D printing had "finally come of age," with two-thirds of all manufacturing companies using the technology in some way. But how has it impacted the AEC industry? It certainly doesn't seem to be transformative, at least at this point.

The major challenges of 3D printing in AEC are material related. Can a material be printed? Is it buildable? Scalability, structural integrity and adherence to codes and regulations are additional hurdles. Finally, site issues often pose a problem.

Forestay's Bruce Bartolf suggests that the tipping point for technologies such as 3D printing may come as a result of the shortage of affordable housing in many parts of the country. "Some kind of automated construction process may push it over the line," says Bartolf. "The housing crisis is creating a big market opportunity for pre-assembled [Accessory Dwelling Units], which are perfect for automating with A.I. and 3D printing. Young people can't afford housing, so that may spark innovation."

On the expectation that 3D printing will undergo significant progress in terms of viability and technology in the next few years, here are some trends and developments we can anticipate:

Increased Speed and Efficiency. Advances in 3D printing technologies could result in increased printing speeds and overall process efficiency. This could make 3D printing more practical for large-scale production, particularly for modular-style building components.

Enhanced Materials. The development of new and improved materials for 3D printing is ongoing. Expect to see a wider range of materials with improved properties, including increased strength, flexibility and resistance to various environmental factors.

Multi-Material and Multi-Color Printing. Flexibility in aesthetics can be a deterrent for 3D printing in design and construction. Future 3D printers may be able to print with an expanded choice of materials or colors in a single job. This could open up new possibilities for creating complex and highly detailed designs and buildings.

3D Printing at the Nanoscale. Researchers are exploring the possibility of 3D printing at the nanoscale, which could lead to the creation of extremely small and precise structures. This has potential applications in features such as smart building components.

Integration with A.I. A.I. could be used to optimize the 3D printing process, from design to printing and post-processing. A.I. algorithms may assist in generating more efficient designs, predicting potential printing issues, and optimizing printing parameters.

Printed Electronics and Wearables. 3D printing is likely to play a role in the production of flexible and wearable electronics. This could lead to the creation of customized and complex electronic devices with integrated functionalities that could help with project management and construction activities.

Advancements in Metal 3D Printing. Metal 3D printing, also known as additive metal manufacturing, is expected to see continued advancements. This includes improvements in printing speed, resolution, and the range of metal alloys that can be printed. Once again, this would provide more options for architects and engineers to consider.

In-Place 3D Printing. The concept of in-place 3D printing involves the ability to print objects directly within the environment where they will be used. This would be a major advancement in the construction industry, overcoming one of the major objections to 3D printing.

3D Printing in Space. Relevant mostly to NASA engineering types, as space exploration and colonization efforts continue, 3D printing is likely to play a crucial role in producing tools, structures, and even habitats in space. This could reduce the need to transport pre-manufactured components from Earth.

Sustainability in 3D Printing. Given the increased focus on sustainable practices, expect to see developments in eco-friendly materials and processes that reduce waste and energy consumption.

3D Printing in Education. The integration of 3D printing in educational settings is likely to grow. This includes the use of 3D printing to teach concepts in science, technology, engineering, and mathematics (STEM) fields. It may also offer productivity improvements to education design and construction projects.

Regulatory Developments. As 3D printing becomes more widespread, regulatory frameworks may evolve to address safety standards, intellectual property concerns, and other legal considerations associated with the technology.

WEARABLES

Wearable technology overlaps with VR/AR in some ways – AR glasses are among the wearable innovations being used by AEC firms. Others include smart boots and helmets, and clip-on devices for measuring, tracking and safety. As with a lot of innovations, wearable technology created a lot of buzz a few years ago, but reality has not met expectations.

Adoption of wearable technology moves ahead, however, so here are some trends and developments we can anticipate in wearable technologies for design and construction firms:

Smart Helmets and Glasses. Wearable devices such as smart helmets and glasses equipped with augmented reality (AR) displays will provide construction workers with real-time information, 3D models, and instructions. This can improve on-site decision-making, accuracy, and efficiency.

AR-Based Design Visualization. Wearables can enable architects and designers to visualize their designs in the real-world environment through AR. This allows for on-site design assessments and collaboration with stakeholders, improving communication and decision-making.

Wearable Sensors for Safety Monitoring. Wearable sensors can monitor vital signs, fatigue levels and environmental conditions, enhancing worker safety. These sensors can provide real-time data to supervisors, allowing for prompt responses to potential health and safety issues.

Exoskeletons for Enhanced Strength and Endurance. We've come a long way from the pulley systems they (probably) used to build the pyramids.

Exoskeletons are wearable devices designed to augment human strength and endurance for lifting and to prevent injury.

GPS-Enabled Wearables for Location Tracking. One of the most commonly used wearables, these GPS-enabled devices aid project management, improve workflow coordination and help ensure that workers are in the right place at the right time.

Wearable Cameras for Documentation. Hello, GoPro. Wearable cameras, whether worn on the head, the body or in glasses, can document construction progress, identify problems and provide a visual record of the construction process and progress for analysis and future reference.

Biometric Authentication for Access Control. Biometric wearables, including fingerprint or facial recognition devices, are being used to secure access control to construction sites.

Haptic Feedback Wearables. A non-verbal communication device that allows users to go beyond visual stimulation with the help of simulated touch, pressure, force, temperature and texture for augmented, mixed, and virtual reality experiences. Wearables with haptic feedback capabilities can provide tactile sensations to users. In construction, this could guide workers or designers through physical spaces based on digital models or design specifications.

Wearable Environmental Sensors. These wearables monitor factors such as air quality, temperature and noise level on construction sites.

Communication and Collaboration Wearables. An example is "hearables," which are more than just smart-headphones that can play music or allow for conversation using Bluetooth. Hearables have additional biometric, movement, or proximity sensors that can help the smart device gather context about the consumer's behavior.

Smart Clothing with Built-in Technology. The integration of technology into workwear, such as vests or jackets, can provide a seamless and non-intrusive way to incorporate sensors, communication devices, and other functionalities.

Training and Skill Enhancement Wearables. Wearables can be used for on-the-job training by providing real-time guidance and instructions to workers. This can be especially useful for new employees or for the introduction of new construction techniques.

Wearable Drones. The Nixie wearable drone, which features a small camera-equipped drone that can be worn as a wrist band, won Intel's $50,000 Make It Wearable competition in 2015. After great initial buzz, the product never seemed to take off (pun intended). Ultimately, however, wearable devices that incorporate drone technology may be used for aerial surveys and inspections.

Integration with BIM. Wearable technologies may integrate with BIM systems, allowing architects, engineers, and construction workers to access and interact with 3D models and project data in real-time on-site.

Flexible and Comfortable Designs. Future wearable technologies will likely prioritize comfort, durability, and adaptability to the rugged conditions of construction sites. This includes considerations of ergonomics and environmental challenges.

DRONES

This section is focused largely on non-AI technologies, but any discussion of drones has to broach the topic of how A.I. can be a game-changer for drones. OpenMind, a non-profit project of Argentinian bank BBVA, offers the following benefits of combining drones and A.I.:

- **Enhanced autonomy.** A.I. algorithms enable drones to operate autonomously, reducing the need for constant human intervention.
- **Improved data processing.** Drones equipped with A.I. can process and analyze data in real-time, making them more efficient for applications like surveillance and monitoring.
- **Adaptive behavior.** A.I. allows drones to adapt to changing conditions and scenarios, making them versatile for various tasks.
- **Increased safety.** A.I. helps drones detect and avoid obstacles, reducing the risk of collisions and accidents.

It really doesn't take too much imagination to think about how a flying drone with the ability to "learn" and "think" could be a far greater asset than one that does neither.

According to the FAA, drones are the fastest growing segment of aviation in the United States. As of October 2023, there were nearly 864,000 drones registered with the FAA, and over 330,000 certified Remote Pilots. These numbers will grow.

The FAA also says that drone use in construction grew by 239% from 2016 to 2021. Here are some trends and developments we can expect to see in drone technology for design and construction in the next few years:

Increased Automation and Autonomy. As noted, drones are likely to become more autonomous, with improved obstacle avoidance and navigation capabilities. This will enable them to perform complex tasks, such as mapping large construction sites, with minimal human intervention.

Extended Flight Times. Advancements in battery technology and energy efficiency are expected to lead to drones with longer flight times. This will allow for more extensive coverage of construction sites and reduce the need for frequent recharging.

AI-Enhanced Image and Data Analysis. Architects and engineers can use drones for tasks such as initial site analysis to project monitoring and documentation.

Site Surveying. Drones equipped with high-resolution cameras and surveying tools can quickly and accurately survey construction sites. This includes capturing topographic data, mapping terrain, and generating detailed site surveys. The data collected aids in the early stages of project planning and design.

3D Mapping and Modeling. Drones create detailed 3D maps and models of construction sites, allowing architects and engineers to visualize the terrain and existing structures more comprehensively.

Aerial Inspections. This was one of the first uses the industry found for drones when they first emerged.

Construction Progress Monitoring. Another common use, drones provide a cost-effective and efficient way to monitor construction progress. Aerial footage and imagery captured by drones can be used to assess the status of the project, track milestones, and identify potential issues that may require attention.

Environmental Analysis. This can be particularly important in difficult, remote or dangerous conditions. Drones equipped with sensors can be used for environmental analysis, including monitoring vegetation, water bodies, and other natural features on or around a construction site. This information is valuable for assessing the impact of a project on the surrounding environment.

Safety Inspections. This is another common use when human access is not desirable. Drones can inspect rooftops, facades, and other elevated structures to identify potential safety hazards.

Marketing and Client Service. Drone footage has become a standard component of presentation to clients, staff and selection committees. Aerial footage captured by drones provides a compelling visual perspective of a project. Running a drone along a newly built bike path or roadway, for example.

Design Visualization. Drones can help with site layout and location.

Volumetric Measurements. Drones equipped with LiDAR or photogrammetry technology can perform volumetric measurements of stockpiles, excavation sites, or other features on construction sites. This data is useful for quantity estimation, resource planning, and project management.

Infrastructure Monitoring. Drones can monitor the condition of infrastructure elements such as roads, bridges, and utility lines. Regular drone inspections can help identify maintenance needs, assess structural integrity, and ensure the safety and reliability of critical infrastructure.

Documentation and As-Built Surveys. Drones can capture comprehensive documentation throughout the construction process. This documentation serves as a valuable record for as-built surveys, providing an accurate representation of the completed project for future reference and maintenance.

Emergency Response and Disaster Assessment. When natural disasters or emergencies occur, drones are being deployed to assess damage, map affected areas, and aid in the planning of emergency response efforts, including for civil and structural engineers, as well as contractors.

Energy Efficiency Assessments. Drones equipped with thermal imaging cameras can assess the energy efficiency of buildings by identifying heat leaks and insulation issues. This information can inform energy-efficient design and retrofitting strategies.

Training and Simulation. Architects and engineers can use drone-captured footage to create training materials and simulations. This is especially beneficial for training construction teams, emergency responders, and other stakeholders on project-specific scenarios.

Customization and Innovation. As drone technology continues to advance, architects and engineers can explore innovative and customized applications. This may include the development of specialized drones for specific tasks or the integration of advanced sensors for enhanced data collection and analysis.

Early on in the drone era, many firms that wanted to use them were stifled by regulations. There are still some areas where local regulations, privacy considerations and safety guidelines prevent the effective use of drones, but these matters appear to be shaking out.

Yet, with potentially new uses for drones in AEC, the issue may arise again. Here are a few ways that drones may continue to advance the industry.

3D Printing Drones. According to MicroCAD Training and Consulting, engineers have designed a team of flying robots with the ability to create structures while hovering in the air. These drones open up the possibility for construction projects to exist in areas that would otherwise be out of reach.

Drone Swarms. Drone swarms can be used to create 3D models of construction sites, which can be used for site planning, design, and analysis. They can also be used to monitor construction progress and measure stockpiles. This would provide substantial value on large construction projects covering a lot of land.

Drone Delivery. We've been hearing for years – and now seeing for real – the prospect of ecommerce delivery using drones. But drones can also be used to deliver materials and equipment to construction sites, reducing the need for manual labor and transportation costs.

Drone Inspection. Drones already inspect and model buildings and infrastructure. As technology improves, the role of drones is likely to expand.

Drone Mapping. Drones can create high-resolution maps and 3D models of construction sites, which can be used for site planning, design, and analysis. Drones have already proven they can map entire cities and create digital twins, and this may become more common in the near future.

Other anticipated advancements with drones include:

AI. Drones equipped with artificial intelligence (AI) algorithms will be able to analyze images and data more intelligently. This includes capabilities such as identifying construction progress, detecting anomalies, and generating actionable insights.

Real-Time Data Transmission. Improved communication systems enable drones to transmit data in real-time. This is particularly valuable for construction project managers who can receive instant updates on site conditions, progress, and potential issues.

Integration with BIM. Drones are increasingly integrating with BIM platforms, allowing AEC professionals to overlay drone-captured data onto 3D models.

LiDAR and Advanced Sensors. Drones equipped with Light Detection and Ranging (LiDAR) and other advanced sensors provide more accurate and detailed mapping of construction sites. This is valuable for terrain modeling, volumetric measurements, and topographic surveys.

Improved Weather Resistance. Drones designed to withstand adverse weather conditions, such as high winds or rain, will become more common. This will increase their reliability and allow for more consistent data collection regardless of the outside environment.

Easier Integration with Ground Robotics. The integration of drones with ground-based robotic systems may become more seamless. This collaboration can enhance the efficiency of tasks such as surveying and inspections, combining aerial and ground-level data.

Regulatory Developments. Regulatory frameworks for drone use in construction is likely to evolve. As drones become more prevalent and sophisticated, governments and aviation authorities may implement rules and standards to ensure the safe and responsible use of drones in construction projects.

Enhanced Thermal Imaging. The emphasis on sustainability is likely to drive the use of drones for thermal imaging. Drones equipped with advanced thermal imaging cameras can detect heat loss in buildings, identify structural issues, and monitor performance of HVAC systems.

Drone as a Service (DaaS) Models. Companies offering Drone as a Service (DaaS) models may become more common. This allows construction firms to access drone technology and expertise without the need for significant upfront investments in hardware, licensing, and training.

DIGITAL TWINS

Digital twin technology holds immense promise across the spectrum of the design and construction industry. From healthcare facilities to water/wastewater pipelines, the benefit of being able to consult a digital representation of a physical asset is nearly immeasurable.

In an article in *Building Design & Construction*, Trimble Customer Experience Team Manager Tiago Ricotta said, "A digital model that nobody accesses or that doesn't provide information doesn't have many purposes. It should be designed in a way that helps gather the information that will help people understand and use the data it provides."

The May 2023 article, "What AEC Professionals Should Know About Digital

Twins," cautions, "For an organization looking to create a digital twin for its building – or AEC firms beginning to build digital twins for their clients – clear objectives should first be set. Without this, it can be difficult to establish what level of detail is needed and whether or not there will be a live link between the model and real life."

The use of digital twins in design and construction is expected to evolve significantly in the next few years, with advancements in technology and greater integration across the industry. Digital twins aid with project visualization, analysis and lifecycle management. Here are some key trends and developments we can expect to see in the use of digital twins in design and construction:

Increased Integration with BIM. Digital twins will become more closely integrated with BIM systems, creating a seamless connection between the design and the operational phases of a project. This integration improves the accuracy and completeness of digital representations throughout the asset's lifecycle.

Real-Time Monitoring and Analysis. This feature is of particular importance to sensitive environments such as clean room and healthcare facilities. Advances in sensors and IoT technologies allow digital twins to provide real-time monitoring of construction sites and built environments. This includes tracking construction progress, monitoring structural health, and analyzing performance data for optimization.

Lifecycle Data Management. Digital twins will be used for comprehensive lifecycle data management, encompassing design, construction, operation, and maintenance phases. This holistic approach ensures that accurate and up-to-date information is available throughout the asset's lifespan.

Energy Performance Optimization. Digital twins will be increasingly used to optimize the energy performance of buildings. This includes monitoring energy usage, identifying inefficiencies, and implementing strategies for energy conservation and sustainability.

AI-Driven Predictive Analytics. One of the primary benefits of digital twins is the ability to foresee issues before they happen based on events that have already transpired.

Advanced Simulation and Visualization. Digital twins offer more advanced simulation and visualization capabilities, allowing stakeholders to explore different scenarios and analyze the impact of design changes before construction begins. This enables more informed decision-making and reduces error risk.

Dynamic Digital Twins. Digital twins will evolve into dynamic models that continuously update and adapt based on real-world conditions.

Digital Twins for Existing Buildings and Infrastructure. This is the business model that many in the AEC and real estate fields dream about – capitalizing on the existing building market with digital twin technology.

Virtual Reality (VR) and Augmented Reality (AR) Integration. Maybe digital twins will be the push the industry needs to finally fully embrace VR/AR. Integrated with AR and VR technologies, digital twins can provide immersive experiences for design reviews, project walkthroughs, and training. This integration enhances collaboration and communication among project stakeholders.

Digital Twin Platforms and Ecosystems. Digital twin platforms and ecosystems will emerge, providing comprehensive solutions that support various aspects of the construction lifecycle. These platforms may include tools for design collaboration, construction monitoring, facility management, and more.

Blockchain for Data Security and Integrity. Blockchain technology may be integrated with digital twins to enhance data security and integrity. Blockchain can provide a decentralized and tamper-resistant system for recording and verifying transactions, ensuring the authenticity of digital twin data.

Robust Data Interoperability Standards. Standards are beginning to emerge for digital twins, but the process has a long way to go. Any lack of standards for data interoperability could be remedied as digital twins evolve and emerge.

Cloud-Based Digital Twins. Cloud-based solutions make digital twins more accessible. This allows stakeholders to access and collaborate on digital twin data from anywhere, promoting greater flexibility and efficiency.

Regulatory Compliance and Documentation. Digital twins can help ensure regulatory compliance by providing accurate and verifiable documentation throughout the construction and operational phases. This is particularly relevant in client markets with stringent regulatory requirements.

SMART BUILDINGS

Unified Micro Systems, an IT solutions provider, reports that the integration of A.I., IoT connectivity, and energy efficiency measures is opening up even greater opportunities for smart building design. The company predicts a surge in innovative technologies like robots, virtual reality (VR), and augmented reality (AR) revolutionizing automated building management in 2024 and beyond.

In a July 2023 article on its website titled "Smart Building Trends in 2023 and What to Look Out For in 2024," the company says, "As smart buildings continue to rapidly evolve, so too do the possibilities for smarter design. In 2024, we can expect to see an increase in the use of biometric security systems and facial recognition technology, as well as voice-activated automation. Additionally, smart materials that are able to respond to their environment will become more common, helping to create buildings that are both efficient and comfortable."

The article predicts that, "The focus on sustainability and energy efficiency will intensify, leading to the development of smart grids that optimize power usage within buildings. As data security concerns grow, the widespread adoption of blockchain applications will ensure the safe storage of sensitive building information, bolstering protection against data breaches.

The smart building platform Sensgreen says that the top seven trends in smart buildings for 2024 are:

Smart HVAC Systems. These systems use sensors and automation to optimize heating, ventilation and A.I.r conditioning systems, ensuring optimal comfort and energy efficiency.

Occupancy Sensors & Space Optimization. These sensors can detect when a room is occupied and adjust lighting, temperature and other settings, reducing energy waste.

Intelligent and Motorized Windows. These windows can adjust their tint and transparency based on the time of day and weather conditions, reducing energy consumption and improving occupant comfort.

Telemetry and Wastewater Treatment. These systems use sensors and automation to monitor water usage and quality, reducing waste and improving sustainability.

Smart Lighting and Motion Sensors. These sensors can detect when a room is occupied and adjust lighting, reducing energy waste.

Biometric Security Systems and Facial Recognition Technology. Sensgreen agrees with United Micro Systems on this. These systems use biometric data to control access to buildings and sensitive areas, improving security and reducing the risk of data breaches.

Voice-Activated Automation. These systems use voice commands to control various building systems, improving occupant comfort and convenience.

Other trends are likely to include regulatory standards, decentralized energy systems and existing building retrofits.

The consensus is that sustainability will remain a focus in smart buildings. We'll also see blockchain technology take a more prominent role, helping to ensure safe storage of sensitive building information and bolstering protection against data breaches.

QUANTUM TECHNOLOGY

IBM says, "Quantum computing is a rapidly emerging technology that harnesses the laws of quantum mechanics to solve problems too complex for classical computers."

According to the website "Interesting Engineering," one of the five greatest breakthroughs in quantum computing occurred in October 2019 when Google confirmed that it had achieved quantum supremacy using its fully programmable 54-qubit processor called Sycamore. "They solved a sampling problem in 200 seconds which would take a supercomputer nearly 10,000 years to solve. This marked a significant achievement in the development of quantum computing."

The May 2023 article, "The 5 Most Significant Breakthroughs in Quantum Computing," concludes, "With a mix of theoretical and experimental demonstrations, quantum computing is still in the early stages of research and development. These developments can potentially revolutionize several industries and academic disciplines, including financial services, materials science, cryptography and artificial intelligence. Even if there is still more study [needed], the implications for quantum computing's future are promising. We may anticipate further developments and innovations in the years to come."

How will quantum computing be helpful to the AEC industry? In short, by speeding everything up...just like it will anything else that involves computing. And while this won't happen tomorrow, it may in the next 5-10 years.

An article on the constructor.org website says, "Quantum computers are super-fast machines that can solve problems beyond the potential of traditional computers by facilitating rapid work, optimizing processes, and improving quality."

While quantum computing is still developing as a useful contributor to the AEC industry, other areas of quantum technology are being used in practice today: Here are a few ways that quantum technology can and will help the AEC industry, according to constructor.org.

Detecting corrosion in structures. Quantum sensors can be used to detect corrosion in structures effectively. These sensors are more sensitive and efficient than existing technologies. Sensors examine small changes in the quality of metals used in construction by studying the variation in magnetism.

Surveying soil. Soil surveys in the initial stage of construction projects provide the information needed for land-use management and land-use planning. Quantum sensors are employed to study the layers below the soil surface. These sensors detect underground abnormalities, disruption, or sinkholes, etc., before the construction begins. Hence, unexpected obstacles are avoided, which helps save time and money caused due to delays.

Better planning. Construction companies rely on the 3D modeling process to enable efficient planning of the construction project. Quantum computing can quickly process complex information to generate a feasible plan for everyone. This can help avoid clashes within different teams and reduce project duration.

Faster project completion. Complex problems can be solved, and solutions can be optimized quickly using quantum computers. The industry relies on quantum computing for planning, surveying, designing, pre-bidding, budgeting, scheduling, quality and cost control, and cost management.

The story concludes, "Quantum computing, together with artificial intelligence, can bring exceptional benefits to the construction industry. However, it is important to note that quantum computing is still in its early stages, and its full potential is yet to be realized."

A DAY IN THE AEC LIFE: 2030

So where will the state of technology be for architects, engineers, environmental scientists and construction professionals when the decade ends? Who knows? But here is a crack at what a day in the life might be like:

COMMUTE TO WORK

The industry and the business world has long realized the value of being together in an office location, so full-time remote work is back to being a rarity. But hybrid is still around, so if it's Monday or Friday, your commute may still be a few steps to the spare bedroom in pajamas. Otherwise, you'll get in your car or SUV and drive – yes, drive; we're still a few years short of fully autonomous vehicles. It is an electric vehicle or at least a hybrid.

Traffic is moving nicely in some places and slowly in others thanks to the recently completed infrastructure improvements that the 2021 Infrastructure Investment & Jobs Act sparked. Thankfully, our legislative and executive branches finally realized how important infrastructure is to the economy and the quality of life.

THE OFFICE

You arrive at the office, wave your watch at the door to unlock it, check in with the company concierge – an A.I.-enabled screen that greets employees and visitors on entry – and give a nod to the Technology Committee that is meeting to review the latest potentially impactful technological innovation that the recent college grad discovered, and figure out ways to capitalize on it. You go to the desk you reserved and set down your much lighter, much sleeker, far more powerful laptop than you had a decade ago. Your smartphone notifies you of a message, which is a holographic image of a recently completed mixed-use building project. The image is a spot-on representation of the structure, though it involved no photographer; it was generated by a prompt using A.I. technology.

The text is from the project technologist and requests a video call to discuss why the smart building's vacuum flush composting toilet isn't working properly. You talk through the matter and solve the problem, but also notice a hot spot on the building's digital twin. You instantly resolve it by adjusting the HVAC system remotely. Wifi9 is a godsend, you think.

You check your watch to see where your team members are, what their utilization rates are for the week, and how they're progressing on the design project and training programs you assigned each of them. You discover that Tony is using an unplanned PTO day, so you instruct the watch to reallocate resources to cover for him on a task you hoped he would do today. It messages two team members with requests to rework their schedules to ensure that the deadline is met.

You also notice it's Mikayla's birthday and tell the watch to wish her happy birthday, which it does. The A.I.-enabled device also remembers that you sent her flowers on her birthday last year and asks if you'd like to do it again. You accept the watch's offer, and though the store is 800 miles away, the process of delivering flowers to Mikayla [by drone, of course] begins.

SITE VISIT

At 10, an alarm bell in the watch rings to let you know it's time for your site visit. You sit down, consult the FAA-approved flight plan you programmed in yesterday, alert one of the company's certified drone pilots and send the device off to visit the site.

MARKETING & BUSINESS DEVELOPMENT – THOUGHT LEADERSHIP

You've scheduled a half-hour to work on the thought leadership/content marketing program that is a pivotal piece of the firm's strategic marketing plan. You open the appropriate A.I. tool from the company's extensive technology software library and provide it with the first of 8 prompts, each based on a lesson learned, interesting project experience or bit of advice that your clients would appreciate.

While you refill your coffee, the A.I. program – MSChatGPT 2030 – produces eight 600-word articles on varying subjects of interest to your clients and peers. You review, edit and then program them to run on the dedicated AEC social media network, twice a week for a month. The review for each story takes about three minutes; it used to take three times longer when you started doing this more than five years ago. Since then, the technology has gotten better, the data it learns from is more robust, and the machine has even managed to capture some of your wit and humor – though it doesn't realize it; it just picked up on some patterns when you fed the completed articles back to it over the years.

SITE DATA REVIEW

You receive a notification that the drone is at the site and transmitting real time data as it hovers above and then traverses along the project boundaries. You voice to your system to open the projection screen so you can see what the drone is seeing. An overlay of the site plan and survey data adds perspective on the project for the eight members of the team who are simultaneously receiving the drone images. A civil engineering subconsultant to the project in Italy notices a conflict and posts it to the chat along with a proposed solution. As team leader, you ask your 3D layout program if the proposed fix is better, and it tells you that it will work. You tell the program to alert the on-site construction manager to the issue and solution, and immediately receive a reply that the message is received and the change will be made.

RECRUITING

It is getting close to 10 a.m. and your wearable technology reminds you that you're scheduled to post a position for a new team member. You instruct your personal computing system to open the HR Recruitment program, then tell it that you want to write a job description for a position very similar to two others in the firm, giving it the names of the people holding those jobs. You add a few different requirements, responsibilities and features, and subtract a few others.

The program produces an ad that begins with a paragraph about the company's culture and what it's like to work there, followed by a brief paragraph about where the role fits in the company, then concludes with the job requirements and responsibilities. You review it, approve it and tell the program to send it along to the dozens of job boards and colleges the firm is subscribed to.

MARKETING & BUSINESS DEVELOPMENT – CRM

You've set aside a half-hour to catch up on industry reading. Your system has organized the stories in your list of online periodicals by prioritizing what it deems to be the most relevant and important. You begin reading and when you encounter the name of a person or firm that you'd like to have in the company CRM system, you touch it. Your A.I. program immediately researches all available data on the subject and either adds it to your contact list, filling out all fields with the data it downloads, or confirming (and correcting, if necessary) the existing entry.

When you're done reading, you're reminded to check through the list of proposed additions to the CRM that A.I. has provided from the messages you received in the last 24 hours. You scroll through the list, swiping right for keepers and left for discards.

PROJECT WALKTHROUGH

It is nearing 11 and it's almost time for the walkthrough you've scheduled with a client for a proposed, but as-yet-unbuilt project. You check your wearable health monitor to make sure you've taken your XR motion sickness pill, which you have. You don your VR glasses with 8K resolution and take a spin through the building to ensure that everything is ready to go. You're mostly concerned that the features you entrusted the A.I. systems to design within your 3D modeling program – in this case stairways and window placement – are thematically consistent, functional and accurate. Not to minimize A.I.'s contribution to the design – it was a huge help with structural analysis, energy efficiency and floorplan layout. It even made some worthwhile suggestions for design ideas based on patterns shown from the firm's previous work. Ultimately, it is a very useful assistant and not a replacement for a skilled professional – at least in 2030.

Once you're satisfied that the virtual building is ready, you contact the client, who has kindly agreed to step away from her family vacation in Oahu for a few minutes. You synch up your VR programs, and while a few other members of the project team watch from their distinct locations, you begin to show the client around.

At one point, the client asks whether it would be possible to move a pillar in one of the rooms to a less conspicuous place. You summon an avatar of the structural engineer to respond to the request and propose options. You ask the program to move the pillar to various locations, but each time a blinking red signal alerts you that there is now a structural weakness in the design.

After a brief discussion, the engineer promises to propose some options in a few days, then disappears from view.

After about an hour of exploring the building, the client thanks you and you make plans for a follow-up virtual meeting. Before she disconnects, she shows you a panoramic view of the Hawaiian beach she's looking at. You're not sure if it's AR or VR, but you're jealous either way.

MARKETING & BUSINESS DEVELOPMENT – RFP RESPONSE

After lunch at the company commissary, you tell your system to call up an RFP you received two days prior. It is very similar to a project you did two years ago, and also to three other projects you proposed on and lost. You tell your system to find the four previous proposals, instructing it which was the win and which were losses. You also tell it to grab the shortlist presentations you did for these projects, and include two pages of notes you made about components of the current RFP project that are unique or distinct from the others, as well as proposed team members, firm members and approved subconsultants.

Within minutes, the system has analyzed the winning and losing proposals and presentations, considered your notes and the requirements of the RFP, and produced a comprehensive 32-page proposal with realistic-looking A.I.-generated design images of the proposed project.

You skim through the output, spotting a few inconsistencies and problems that you'll need to fix in the two weeks between now and when the proposal is due. But it's getting toward 2 p.m. and you decide that you'll finish revieing the proposal in your home office. Your wearable reminds you that the company's commitment to the health of its employees recommends a gym visit, so you'll squeeze that in on the way home.

Before you close up shop you review and lightly edit the three preprogrammed project progress reports that your A.I. assistant drafted for three separate clients based on the tasks you completed and the project's scope and schedule. You send the progress reports, tell your A.I. assistant to update your workday status, grab your laptop and head to the parking lot.

CHAPTER 6 TAKEAWAYS

- Tech-forward firms perform better than their tech-static counterparts.
- Tech forward firms embrace and adopt technology to improve systems and processes.
- Open-Source Systems are a growing presence in AEC firms. They are typically less expensive and more user-friendly than other technologies.
- Virtual Reality/Augmented Reality have yet to meet their promise in the AEC industry, but a breakthrough may be on the horizon.
- 3D printing has also not met its lofty expectations, but it is currently a useful tool with the potential to expand its adoption in the AEC industry.
- Wearable technology has many variations and uses, and it is another growing segment for AEC firms.
- Drone use has exploded in the nation and the industry, and the growth shows no signs of subsiding, especially as machine learning and other A.I. technologies take hold.

- Digital twins offer myriad uses in the industry, and is another emerging technology expected to play a much larger role in years to come.
- Smart building technology will grow in sophistication and use.
- Quantum computing has not yet reached the point where it is affecting the AEC industry, but it will be huge when it comes. In the meantime, existing quantum technologies are making a difference today.

About the Author

Jerry Guerra is a writer, editor and 30-year consultant to the AEC industry. He has authored more than a dozen AEC Industry books, including several published by PSMJ. Jerry has addressed the national conventions of the AIA, ACEC and World Trade Center Montreal.

He has written and ghostwritten hundreds of articles for dozens of industry and mainstream publications, including Engineering-News Record, the Boston Business Journal, Lab Manager, School Planning & Management, Roads & Bridges, Public Works and Architecture magazine. He has also been quoted in the Wall Street Journal and many other publications.

The Design-Build Institute of America honored Jerry with its Press Award for Excellence in Reporting on Design-Build Issues.

In 2002, Jerry founded The JAGG Group, a management consulting firm based in Massachusetts. The JAGG Group focuses on providing marketing, market research, and public relations services to firms in the architecture, engineering, environmental, and construction industries. The firm is affiliated with a network of consultants who provide full-spectrum management consulting services, including merger and acquisition, business planning, recruiting, and financial services.

Prior to becoming an editor and consultant to AEC firms, Jerry was a sportswriter, news writer, editor, and bureau chief for daily newspapers.

Jerry graduated from the University of Rhode Island with a bachelor's degree in English and History. He can be reached at jguerra@jagg-group.com. Check out his AEC Insight blog at jagg-group.com.

REFERENCES

CHAPTER 1

Brittanica, "Artificial Intelligence," by B.J. Copeland
https://www.britannica.com/technology/artificial-intelligence

Simplilearn, "What is Artificial Intelligence and Why It Matters"
https://www.simplilearn.com/tutorials/artificial-intelligence-tutorial/what-is-artificial-intelligence

IBM, "What is a Chatbot"
https://www.ibm.com/topics/chatbots#:~:text=A%20chatbot%20is%20a%20computer,conversation%20with%20an%20end%20user.

Zendesk, "Deep vs. Machine Learning," by Patrick Grieve, November 14, 2023
https://www.zendesk.com/blog/machine-learning-and-deep-learning/

Amazon Web Services, "What's the Difference Between Machine Learning & Deep Learning?"
https://aws.amazon.com/compare/the-difference-between-machine-learning-and-deep-learning

Zebra Networks, "What is the Difference Between Machine Vision & Computer Vision?"
https://www.zebra.com/us/en/resource-library/faq/what-is-the-difference-between-machine-vision-computer-vision.html

Time, "When Might AI Outsmart Us? It Depends Who You Ask," by Will Henshall, January 19, 2024
https://time.com/6556168/when-ai-outsmart-humans/

Lifewire.com, "Strong AI vs. Weak AI: What's the Difference?" by Robert Earl Wells III, June 15, 2023
https://www.lifewire.com/strong-ai-vs-weak-ai-7508012

Marketing Artificial Intelligence Institute, "How to Cut Through the Hype About AI," April 27, 2020
https://www.marketingaiinstitute.com/blog/how-to-cut-through-the-hype-about-artificial-intelligence-in-marketing

Forbes, "Artificial Intelligence is Not a Technology," by Kathleen Walch, May 1, 2018
https://www.forbes.com/sites/cognitiveworld/2018/11/01/artificial-intelligence-is-not-a-technology/?sh=63f0e9b31d1d

Tech Target, "Turing Test," by Benjamin George & Alan S. Gillis, April 2023, https://www.techtarget.com/searchenterpriseai/definition/Turing-test

Model Evaluation & Threat Research, "2023 Year in Review"
https://metr.org/blog/2024-02-07-2023-year-in-review/

The White House, "Fact Sheet: P resident Biden Issues Executive Order on Safe, Secure, and Trustworthy Artificial Intelligence," October 30, 2023
https://www.whitehouse.gov/briefing-room/statements-releases/2023/10/30/fact-sheet-president-biden-issues-executive-order-on-safe-secure-and-trustworthy-artificial-intelligence/

CHAPTER 2

AIA/Deltek Architecture Billing Index (ABI)
https://www.aia.org/resource-center/aiadeltek-architecture-billings-index

Practice in Public, "These Words Make It Obvious That Your Text is Written by AI," by James Presbitero, Jr., December 31, 2023
https://medium.com/practice-in-public/these-words-make-it-obvious-that-your-text-is-written-by-ai-9b04f399d88c

The Startup, "There are No Special Words That Give Away AI-Generated Content," by Toby McInnis, January 22, 2024
https://medium.com/swlh/there-are-no-special-words-that-give-away-ai-generated-content-d8781a689422

Content Marketing Institute (CMI), ""How to Put A.I. to Work for Better Content Marketing," by Jodi Harris, August 31, 2023
https://contentmarketinginstitute.com/articles/ai-content-operations-plan/#:~:text=First%2C%20make%20sure%20incorporating%20AI,your%20marketing%20and%20business%20priorities.

Technopedia.com, "The Five Best Cover Letter Generators Compared for 2024," by Krishi Chowdhary, February 29, 2024
https://www.techopedia.com/ai/best-ai-cover-letter-generator

The Generator, "Midjourney 6 Means the End to a Big Chunk of the Photo Industry," by Thomas Smith, January 22, 2024
https://medium.com/the-generator/midjourney-6-means-the-end-for-a-big-chunk-of-the-photo-industry-068cb5faeddc

DDI Development, "AI in CRM: Benefits and Use Cases of AI-Powered CRM System," December 2023
https://ddi-dev.com/blog/programming/ai-in-crm-benefits-and-use-cases-of-ai-powered-crm-system/

Workable, "Seven Ways AI is Used in Human Resources," by Keith McKenzie, December 2023
https://resources.workable.com/tutorial/artificial-intelligence-in-human-resources#7-ways-ai-is-used-in-human-resources

Deloitte, "2024 Engineering & Construction Industry Outlook," by Michelle Meisels, Misha Nikulin, Kate Hardin, Matt Sloane and Kruttika Dwivedi
https://www2.deloitte.com/us/en/insights/industry/engineering-and-construction/engineering-and-construction-industry-outlook.html

New Jersey Society of Certified Public Accountants (NJCPA), "The AI Revolution: Transforming Accountants' Roles," by Sean P. Breheney, September 12, 2023
https://www.njcpa.org/article/2023/09/12/the-ai-revolution-transforming-accountants-roles#:~:text=By%20automating%20compliance%20processes%2C%20AI,value%20in%20other%20practice%20areas.

Architizer, "15 Top AI Tools for Architects and Designers," by Samantha Frew
https://architizer.com/blog/practice/tools/top-ai-tools-for-architects-and-designers/

McKinsey, "Reinventing Construction Through a Productivity Revolution," by Filipe Barbosa, Lola Woetzel, Jan Mischke, Maria João Ribeirinho, Mukund Sridhar, Matthew Parsons, Nick Bertram and Stephanie Brown, February 27, 2017
https://www.mckinsey.com/capabilities/operations/our-insights/reinventing-construction-through-a-productivity-revolution

CHAPTER 3

MIT Tech Review, ""Google DeepMind's New Gemini Model Looks Amazing – But Could Signal Peak A.I. Hype," by Melissa Heikkilä & Will Douglas Heaven, December 6, 2023
https://www.technologyreview.com/2023/12/06/1084471/google-deepminds-new-gemini-model-looks-amazing-but-could-signal-peak-ai-hype/

Tech Target, "6 Top Predictive Analytics Tools for 2024," by George Lawton, January 8, 2024
https://www.techtarget.com/searchbusinessanalytics/tip/6-top-predictive-analytics-tools

The Friedman File, "The AI Dilemma: How Artificial Intelligence Can Help You (and Possibly Hurt You), by Richard Friedman, June 2023
https://friedmanpartners.com/the-ai-dilemma-how-artificial-intelligence-can-help-you-and-possibly-hurt-you/

Autodesk Blog, "Evolving Your AEC Toolbox with Autodesk AI," by Amy Bunszel, November 13, 2023
https://www.autodesk.com/blogs/aec/2023/11/13/ai-in-aec/

AEC Magazine, "Autodesk: Laying the Foundation for AI," by Martin Day & Greg Corke, December 4, 2023
https://aecmag.com/ai/autodesk-laying-the-foundations-for-ai/

Technology, ""Bentley Systems Turns A.I. Into an Infrastructure Assistant," by Tom Swallow, December 4, 2023
https://technologymagazine.com/articles/bentley-systems-turns-ai-into-an-infrastructure-assistant

FutureIoT, "Bentley Systems: 2023 is the Start of Infrastructure Intelligence," by FutureIoT Editors, November 2, 2023
https://futureiot.tech/bentley-systems-2023-is-the-start-of-infrastructure-intelligence

CHAPTER 4

Unanet AEC Inspire 2022-2023
https://info.unanet.com/inspire-report-2022-23

MIT Sloan, "How Generative AI Can Boost Highly Skilled Workers' Productivity," by Meredith Somers, October 19, 2023
https://mitsloan.mit.edu/ideas-made-to-matter/how-generative-ai-can-boost-highly-skilled-workers-productivity#:~:text=A%20new%20study%20on%20the,who%20don't%20use%20it.

Deltek Blog, "Bridging the Gap Between A.I. and Business Value, by Elin Hauge, November 2023
https://www.deltek.com/sv-se/blog/ai-business-value

PSMJ, 2024 AEC Artificial Intelligence Policies, January 2024
https://go.psmj.com/2024-ae-artificial-intelligence-policies

Crowdstrike, "Generative AI (Genai) in Cybersecurity," by Lucia Stanham, November 27, 2023
https://www.crowdstrike.com/cybersecurity-101/secops/generative-ai

CHAPTER 5

Oxford Martin, "The Future of Employment: How Susceptible are Jobs to Computerisation?" by Carl Benedikt Frey & Michael Osborne, September 1, 2013
https://www.oxfordmartin.ox.ac.uk/publications/the-future-of-employment/

"Will Robots Take My Job"
https://willrobotstakemyjob.com/

B2B News, "What Are the Greatest Threats to Small Business From A.I.?" by Adam Tanton, July 17, 2023
https://www.b2bnn.com/2023/07/what-are-the-greatest-threats-to-small-businesses-from-ai/

Forbes, "10 Hurdles Companies are Facing When Implementing A.I.," by Brian David Crane, October 25, 2023
https://www.forbes.com/sites/theyec/2023/10/25/10-hurdles-companies-are-facing-when-implementing-ai-and-how-to-overcome-them/?sh=c4962474c91c

Small Business & Entrepreneurship Council, "Small Business Adoption Survey," October 2023
https://sbecouncil.org/wp-content/uploads/2023/10/SBE-Small-Business-AI-Survey-Oct-2023-FINAL.pdf

Center for Strategic & International Studies, "AI Regulation is Coming: What is the Likely Outcome," Bill Whyman, October 2023
https://www.csis.org/blogs/strategic-technologies-blog/ai-regulation-coming-what-likely-outcome

MIT Sloan Management Review, "Using Federated Machine Learning to Overcome the AI Scale Disadvantage," by Yannick Bammens and Paul Hünermund, August 22, 2023
https://sloanreview.mit.edu/article/using-federated-machine-learning-to-overcome-the-ai-scale-disadvantage/

Forbes, "The 10 Best Examples Of Low-Code And No-Code A.I.," by Bernard Marr, December 12, 2022
https://www.forbes.com/sites/bernardmarr/2022/12/12/the-10-best-examples-of-low-code-and-no-code-ai/?sh=3e92adae74b5

The Generator, "My 9 GenerativeAI Predictions for 2024," by Thomas Smith, December 19, 2023
https://medium.com/the-generator/my-9-generative-ai-predictions-for-2024-a3a8767b00ad

Educause Review, "7 Things You Should Know About A.I.," December 6, 2023
https://er.educause.edu/articles/2023/12/7-things-you-should-know-about-generative-ai

Bipartisan Policy Center, "Navigating the Era of A.I.: Implications for Small Businesses," by Michelle Kumar, November 3, 2023
https://bipartisanpolicy.org/blog/navigating-the-era-of-ai-implications-for-small-businesses/

Bloomberg, "AI Legal Protections May Not Save You From Getting Sued," by Brad Stone, November 13, 2023
https://www.bloomberg.com/news/newsletters/2023-11-13/legal-protections-from-ai-companies-may-not-save-you-from-getting-sued?embedded-checkout=true

Harvard Business Review, "Generative A.I. Has an Intellectual Property Problem," by Gil Appel, Juliana Neelbauer, and David A. Schweidel, April 7, 2023
https://hbr.org/2023/04/generative-ai-has-an-intellectual-property-problem

CHAPTER 6

Unanet Blog, "What Gives AEC Firms an Edge in the War for Talent," by Lucas Hayden, December 13, 2023
https://unanet.com/blog/what-gives-aec-firms-an-edge-in-the-war-for-talent

Synopsys, "What is Open Source Software & How Does It work
https://www.synopsys.com/glossary/what-is-open-source-software.html#:~:text=Definition,distribution%20with%20its%20original%20rights.

Gartner, "3 Reasons Why VR and AR Are Slow to Take Off," by Christy Pettey, September 6, 2018
https://www.gartner.com/smarterwithgartner/3-reasons-why-vr-and-ar-are-slow-to-take-off

"Augmented Reality Application Areas for the Architecture, Engineering, and Construction Industry," by Sara Rankohi, Mahsa Rezvani, Lloyd Waugh and Zhen Lei, September 15, 2023
https://www.intechopen.com/chapters/1152194

AEO Design & Construction, "The Future of AEC with VR and AR," by Kamaljeet S. Marwah, December 6, 2023
https://www.aeodc.com/the-future-of-aec-with-vr-and-ar

Harvard Business Review, "The 3D Printing Revolution," by Richard D'Avini, May 2015
https://hbr.org/2015/05/the-3-d-printing-revolution

PwC, "3D Printing Comes of Age in US Industrial Manufacturing,"
https://www.pwc.com/us/en/industries/industrial-products/library/3d-printing-comes-of-age.html#:~:text=It's%20become%20clear%20that%20the,a%20printed%20NASA%20rocket%20engine.

OpenMind, "AI and Drones," by Ahmed Banafa, December 11, 2023
https://www.bbvaopenmind.com/en/technology/artificial-intelligence/ai-and-drones/

Federal Aviation Administration (FAA), "Drone Safety Day," February 27, 2024
https://www.faa.gov/uas/events/drone_safety_day

MicroCAD Training & Consulting, " Key Construction Trends of 2024 – Unleashing the Power of AR and Drones," by Sarah Redick, January 25, 2024
https://microcad3d.com/construction-2024-with-ar-drones-aec-collection

Building Design & Construction, "What AEC Professionals Should Know About Digital Twins," by Quinn Purcell, May 8, 2023
https://www.bdcnetwork.com/what-aec-professionals-should-know-about-digital-twins

Unified Micro Systems, "Smart Building Trends in 2023 and What to Look Out For in 2024," July 2023
https://www.unifiedmicro.systems/articles/smart-building-trends-this-year

IBM, "What is Quantum Computing"
https://www.ibm.com/topics/quantum-computing

Interesting Engineering, "The 5 Most Significant Breakthroughs in Quantum Computing," by Tejasri Gururaj, May 11, 2023
https://interestingengineering.com/science/5-breakthroughs-in-quantum-computing

Constructor.org, "How Can [the] Construction Industry Utilize Quantum Computing"
https://theconstructor.org/artificial-intelligence/quantum-computing-construction-indus/554265/#google_vignette

PSMJ | Resources, Inc.®

Made in the USA
Las Vegas, NV
16 April 2026

45387858R00125